Waterworks in the Netherlands

Tradition and Innovation

Mills and Pumping Stations, Kinderdijk

Storm Surge Barrier, Oosterschelde

Lely Pumping Station, Medemblik

Koning Willem-Alexander Pumping Station, Katwijk aan Zee

Osen Lock, Osen

R.J. Clevering Scouring Sluice, Lauwersoog

Waterway Forest, Marknesse

Waterworks in the Netherlands

Tradition and Innovation

Concept and photography Luuk Kramer

Texts by Theo van Oeffelt and Bernard Hulsman

With contributions by Inge Bobbink and Ellen Vreenengoor

Editors Inge Bobbink, Eric Luiten and Lodewijk van Nieuwenhuijze

Illustrations Emma Ottevanger and Inge Bobbink

Luuk Kramer

nai010 publishers

Contents

Preface

Innovative, impressive, and iconic: that is how the Dutch waterworks can best be described. In other words, these are the instruments – the civil engineering structures (or 'kunstwerken', works of art, in Dutch) – that regulate the water. Unlike dikes, which block the water, these waterworks have a button or a lever: they are locks, weirs, pumping stations, culverts, and watermills. In a complex game of cooperation, they ensure that we keep our feet dry in the Netherlands, and that the water level can almost always be kept constant in both the water-rich western part of the country, as well in as the drier and higher eastern part.

Over many thousands of years, a delta landscape in which several large rivers flow into the sea has transformed into the intensively inhabited and cultivated country of the Netherlands. In that period, the country has established a long tradition of defending itself against, and controlling, the sea and river waters. The first waterworks were built almost two centuries before Christ, well before the arrival of the Romans. Small dams in which a small wooden tube with a valve was placed made it possible to have the water to flow in one direction, thereby draining the land, and protecting it from outer water. The Romans then made some more drastic changes in the water management. In particular, the Canal of Drusus, an excavated connection between the IJssel and the Lower Rhine, still captures the imagination. Gradually, our ancestors again came to take the initiative themselves, which was a development that was also forced by the population growth that occurred around the first millennium change. In villages of all kinds, inhabitants began working together to the extract peat soil and make it suitable for agriculture or housing. Free peasants and farmers had an equal interest; one was both the owner as well as the user of the land. Changing property relationships, for example those acquired through leasing, led to the unification of village administrators in an intensive partnership: the water boards. In 1255, Count William II of Holland founded the first official water board: the Hoogheemraadschap van Rijnland (the Rheinland High Water Board), which still exists today, albeit in a slightly different form.

Yet the water boards were unable to combat the frequent flooding in a coordinated way, which was the reason for the creation of the Bureau voor den Waterstaat in 1798. The main tasks of this predecessor to Rijkswaterstaat, as it was renamed in 1848, were the management and the maintenance of the rivers, and the supervision of the water boards. The mission of the water boards was later enshrined constitutionally in 1848: the water boards were responsible for both the quantity and (these days) the quality of the water.

Over the next 100 years, until about the middle of the twentieth century, the developments in hydraulic engineering were still first and foremost in the hands of the water boards and the local and regional authorities. For example, the Netherlands had 3,500 water boards in 1850, a number that has since fallen sharply to 22 in 2017. At the national level, the government mainly played a coordinating and regulatory role. But after the flood disaster of 1953, this changed drastically, with the political ratification of the Delta Act in 1957, followed four years later by the Delta Plan. The government took control, with Rijkswaterstaat as its executive body. The commissioning authority for the large breakwaters (dikes) and waterworks shifted from a decentralised to a centralized structure. Rijkswaterstaat became the most prominent builder.
A second turnaround followed a few decades later. The planning that until then had always been technical in nature, and which focused on providing resistance to the water, transformed in the 1990s into a reflection on the spatial component, and on resilience: dealing with the water in a different way. The battle against the water gave way to 'working together with water' and 'building with nature,' whereby natural processes are used as a means of protecting against the water, for example by activating sedimentation and marsh formation, with the goal of lowering water levels off the coast, and reducing the force of storm surges.

Another such turnaround is currently at hand: that towards a structure with multiple commissioning bodies, and a more multi-functional mission for waterworks, for example one that includes energy-generating components. In the thinking about commissioning, management, and functioning, a wave movement can now be detected that is comparable to the relationship between the public and private domains, as Auke van der Woud sketched in **Het lege land** for the first half of the nineteenth century.

Prologue

The Netherlands – to be more precise, the lower part of the Netherlands – was largely formed by the Rhine and Meuse rivers, which culminate in our country. Their mouths, which have also been widened into estuaries by tidal action, are nowadays located in the provinces of Zeeland and Zuid-Holland. The Betuwe, the Bommelerwaard, and parts of the Groene Hart (Green Heart) were originally part of the sphere of influence of these strongly branching rivers. One of the northern Rhine branches, the Old Rhine, silted up in the eleventh century, and lost its access to the sea. The Rhine branches of the IJssel and the Vecht culminate in the former Zuiderzee, which was closed off by the construction of the Afsluitdijk in the 1930s. Other river branches, cut off from the major rivers in the current context, once filled the swampy lagoon behind the Hollandish beach ridges. This created large peatlands: land whose elevation increased due to the accumulation of plant material. Peat rivers carried the excess water to the large rivers. Through human intervention, namely the making habitable of the peat marshes, peat was mined as far back as the seventh century A.D. To this end, ditches were dug at right angles to the peat rivers, which caused the soft soil to settle so deeply over several centuries that it dropped below the river's high-water level. Small dikes were constructed to separate the inland water from the outer water[1]. When the outer water levels were low, the first waterworks were capable of carrying the water from the polder to the low-lying, diked land. Population growth led to an ever-increasing intensification of land cultivation, which in turn improved water regulation more and more; the development of hydraulic engineering knowledge, as a result of the increasing number of waterworks, made this possible. For example, structures were built to dam the water in the river, which improved navigability. New works were built to turn the water, and thereby divert the watercourse in a different direction, or to temporarily block the water and then sluice it through when the outer water was at a low level. In the ninth century, waterworks were developed that could pump the water upwards. At first this was done using human and horsepower, and later, from the eleventh century onwards, by using the wind. The development of steam, diesel, and electric motors led to the construction of weirs and pumping stations that have become an inextricable part of the polder landscape.
These many thousands of weirs, pumping stations, and locks are still operated by the water boards. Dijkgraaf Roelof Bleker of the Rivierenland Water Board shows the control room in Tiel, from where about 180 local pumping stations and 340 local weirs are operated using telemetry, from very small pumps to the screw-pumping mills at Kinderdijk. A further 1,700 weirs are controlled manually.
Roelof Bleker has been a dijkgraaf since 2010. When he took office, he said: 'I think it will be fantastic to work together with other parties to give more space to water, to nature, to safe dikes, and to the economic reinforcement of our beautiful area. I am also looking forward to the campaign to preserve the water boards for our country.'
Now, seven years later, he says about that last point: 'We have fought this battle, and it no longer has the relevance that it had then. There are still 22 water boards left; a few of them may still merge, but then it will stay that way. The recognition of our work is clearly embedded in both society and politics.'
The Water Board has certainly worked hard on this embedding process in recent years. The Waterschap Rivierenland has developed a great deal of expertise on themes such as sustainability and the energy transition, always in close cooperation with many and often very different partners, in addition to the 'natural partners' such as municipalities, as well as the business community and housing corporations, for example.

Henk Ovink is the (first) Dutch water envoy. When he was appointed in 2015, he said that 'worldwide, water is the binding challenge. The Netherlands, with its integrated approach to water scarcity, flood risk management, water quality, logistics, and transport all over the world, makes the difference. We are the frontrunner, and that suits us well, but we are also obliged to invest and to innovate, both within the Netherlands and abroad. The Dutch knowledge and skills are unique, and I see plenty of opportunities to make even better use of them internationally.'

This international appreciation was also apparent during the interview for this essay, when **the New York Times** sent Ovink the draft text for a large article about the Dutch approach to water (published on 15 June 2017, under the title **The Dutch Have Solutions to Rising Seas – The World is Watching**). 'Like cheese in France or cars in Germany, climate change is a business in the Netherlands,' the newspaper wrote.
As Ovink says, 'We have developed a centuries-old tradition that combines safety and quality. In addition to managing the risks of water damage, we have added beauty as a value. Functional and beautiful: we have been succeeding in this combination for many centuries. For example, look at the Beemster, which for good reason has been designated as a World Heritage site.'

That long tradition of defending against, and controlling, the sea and river water has resulted in countless waterworks. Built constructions that span a wide range in terms of function, material, size, appearance, and location. Most of the waterworks are used to nudge the water flows in the desired trajectories and directions, and to protect the country from an overflow of water. They regulate the flow volumes and the direction of natural waters such as seas, lakes, rivers, and streams, as well as the excavated surface water: the ponds, canals, channels, and ditches.

Photography
The waterworks (civil engineering structures) photographed for **Waterworks in the Netherlands** are mainly situated in the following types of landscape: coastal and marine clay areas, the peat meadowland and reclaimed land areas, and the transitional areas between these landscapes (p. 26, map waterworks). This is not so surprising, considering that these landscape types in particular derive their identity from the transformation of the Delta. In addition, for most of these projects, illustrations have been added that further explain their location in the landscape, and the operation of the waterwork.

Reflections
Bernard Hulsman's architectural reflections on the photographed waterworks show that their design is not only determined by a functional programme of requirements. 'A bicycle shed is a building; Lincoln Cathedral is a piece of architecture,' wrote British architecture historian Nikolaus Pevsner in **An Outline of European Architecture, 1943**. To be called architecture, a building must have been designed with an aesthetic purpose: that is how he explained the difference between the bicycle shed and the cathedral in Lincoln. In other words, a bicycle shed is purely functional, in the sense of being efficient, and is therefore a building, whereas a cathedral is first and foremost meant to be 'beautiful,' and is therefore architecture.
At first sight, the Dutch waterworks are like Pevsner's bicycle sheds. If there is one type of structure that needs to be functional, it is a waterwork. After all, without waterworks, a large part of the Netherlands would not exist. Nevertheless, waterworks in the Netherlands are called 'kunstwerken', or 'works of art' (versus 'civil engineering structures' in English). And rightly so. The Dutch waterworks are not merely simple and purely efficient constructions, but instead are (much like the bicycle parking facilities in today's Netherlands, incidentally) indeed architecture.

1. The outer water includes rivers with an open connection to the North Sea and the Wadden Sea.

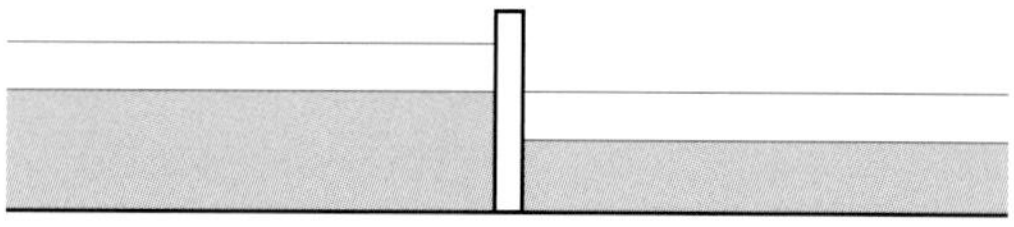

Fixed weir

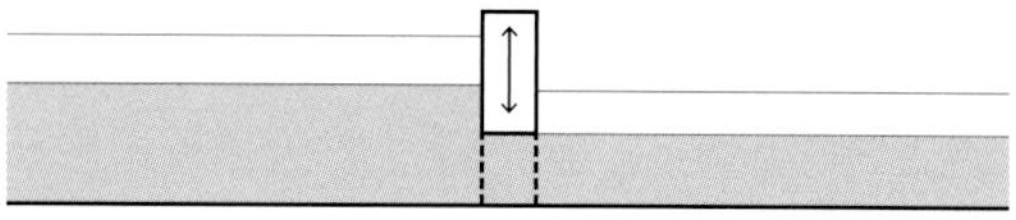

Movable weir

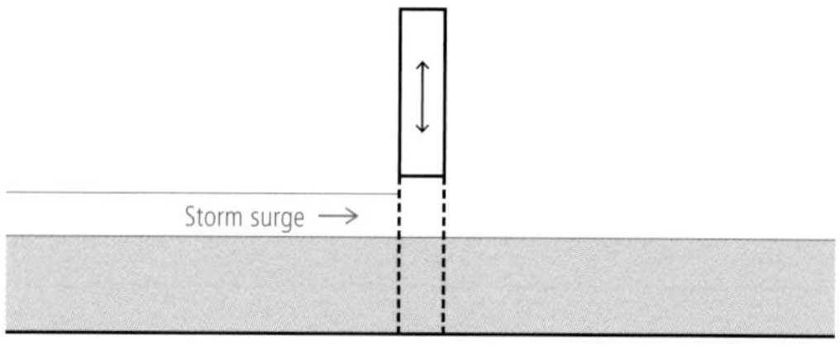

Barrier

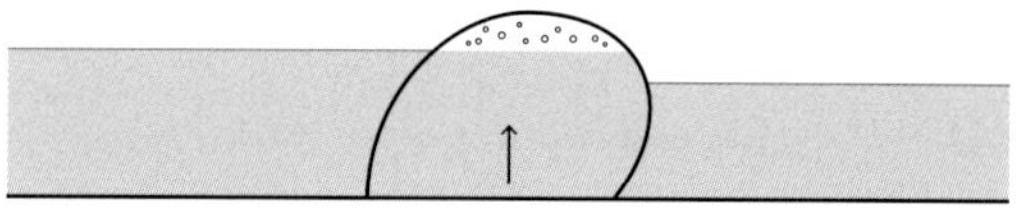

Inflatable rubber weir

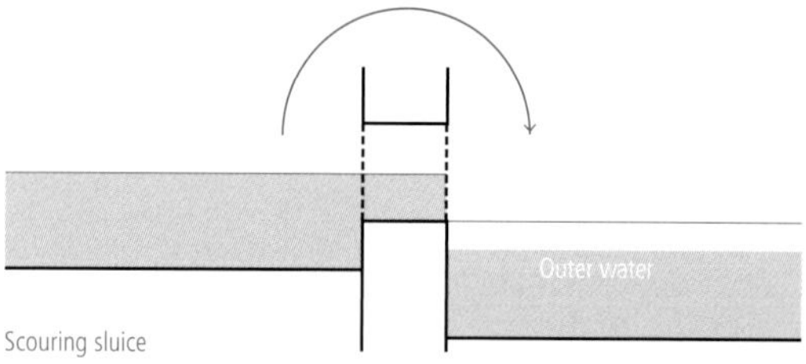

Scouring sluice

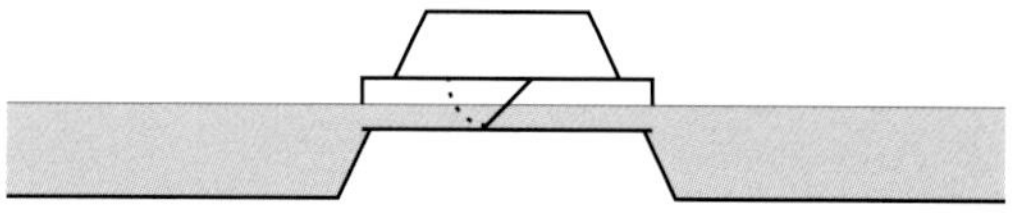

Culvert valve

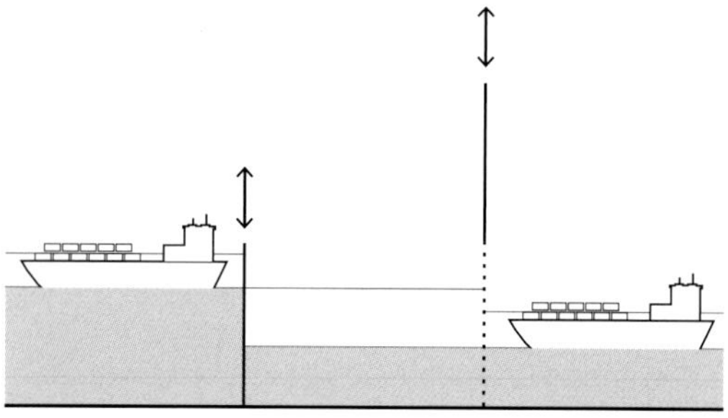

Navigation lock

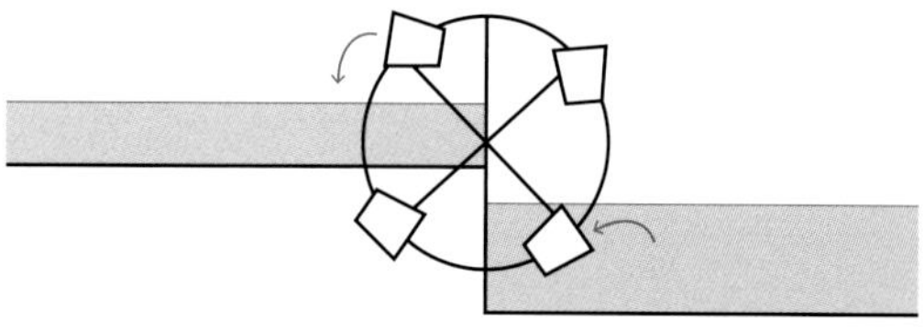

Waterwheel

Technique

Within this book, **three types of waterworks** are distinguished: the weir and the barrier, the lock, and the pumping station with its predecessor, the mill. The large monumental weirs and barriers are mainly located in the coastal and marine clay areas (the landscape that has been strongly influenced by the dynamics of the sea), and in the region of the big rivers passing from the southeast to the west of the Netherlands. These waterworks were constructed to protect the land from sea flooding, and to tame the natural flow of water from the rivers. These rivers are 'tamed' by dividing them into smaller sections, whereby the water comes to rest at more or less the same level. This allows for sailing in both directions throughout the year. Pumping stations, and previously mills, are inextricably linked to the peat meadowland and the reclaimed land areas. This type of landscape was created by the reclamation of the former wetland areas. Locks, on the other hand, are situated in every type of landscape, because of their mediating function between the man-made differences in height that exist between various watercourses.

Weirs and barriers are fixed or movable constructions that hold back the water, in the transverse direction to the waterline, causing differences in water levels on both sides of the weir or barrier. There are fixed weirs, which sometimes consist only of a wooden plank in a ditch, and large variable weirs such as the Oosterscheldkering (Oosterschelde storm surge barrier) pp. 126-131. With a fixed weir, there is a different water level on either side of the element. Movable weirs can set a water level difference as circumstances demand, such as changing water volumes depending on the season. If a weir has to hold back the water on only one side, then it is called a barrier. The Oosterscheldkering holds back the seawater in the event of a storm, so that the water level in the Oosterschelde does not rise too high, which could pose a threat to the dikes and the land behind it. Weirs and barriers are especially visible when they are movable, and also when they are large, such as weirs located in rivers, or between a river and the sea. A weir in a river can hang like a bridge element above the river, or can slide down or fold back into the watercourse. This is the case, for example, at the weirs near Grave pp. 110-113 and Waalwijk pp. 42-45. The same sliding mechanism is used for the barrier at Krimpen aan den IJssel pp. 114-119, where the slider can hold back the saltwater when the seawater rises to dangerous levels. The water then cannot penetrate any further inland, and the polders remain free from flooding. The Maeslantkering pp. 132-137 is very visible and striking, especially when the gigantic, semi-concentric sliders rest on the riverbank. Here, too, the seawater must be held back in the event of a storm. The inflatable weir at Kampen pp. 142-145 has a completely different mechanism: the weir, made of rubber fabric, lies on the bottom of the river between the Zwarte Meer and the Ketelmeer, and is inflated from the IJsselmeer when high waters threaten. Seen in this context, the inflatable weir is actually a barrier. In de Lek, there are three large, striking, movable, and curved weirs. The weir at Driel pp. 120-125 is closed at normal and low water levels. The water is divided between the Lower Rhine and the river Lek, allowing smooth and safe shipping to remain possible. The weirs at Amerongen and Hagestein are situated further downstream in the river Lek, and are only opened during high water levels, so that the water can be quickly discharged to the sea. Barriers – and in particular weirs that are located in navigable waters – are often combined with navigation locks, because the difference in height between the two water levels has to be overcome on both sides of the closed weir.

Locks consist of movable sliders or doors, in addition to a fixed section that covers the river bottom and the banks of the water course. Locks block the water just like a dam, a barrier, or a weir, but only when and for as long as necessary. The actual goal is to ultimately discharge, drain, or sluice away the water. The drainage lock or scouring sluice is the successor to the culvert valve, which usually lies invisible in the dike. In the Middle Ages, the first locks with wooden doors were introduced when people began building dikes and quays on a large scale. The lock was able to discharge larger quantities of excess water via a single point, which made many culvert valves superfluous. A scouring sluice can regulate the water outflow in such a way that the stream of sludge (which can accumulate in a harbour basin, for example) can be washed away. There are also locks that can be opened to temporarily allow water from the basin to flow into a designated and prepared polder so that

the basin's storage capacity is relieved. A nice example is the inlet at the Rotte river pp. 28-31 where water can be temporarily fed into the Eendrachtspolder, which serves as an emergency polder. As soon as the Rotte has returned to its normal level, the water from the polder can be discharged via the regular pumping station.
The navigation lock is a further development of the simple lock, probably built in the Netherlands for the first time in the fourteenth century. A navigation lock has two gates or sliders, unlike a 'normal' lock. The kolk (chamber), as the water between the two doors is called, can transport a boat or several ships in a controlled environment, from one water level to another, by means of letting water in and out. Depending on how the closing door moves, we can distinguish between several different types of navigation lock. Hinge gates, pointed gates, and spindle gates are among the rotary closing systems, which move along a vertical axis. Those with pointed gates pp. 28-31, originally made of wood and now mainly made of steel or composite, were and continue to be widely used.
Navigation locks are located between waters with a difference in level: between a lake at polder level and a basin[II], between a canal and a river, or within a river if it has been dammed, such as the lock and weir at Grave. Large modern locks on heavily trafficked canals are activated by means of sliders that hang in between large portals, and which can be quickly moved up and down. A good example of this is the Prins Bernhardsluizen at Tiel pp. 76-79, which connect the Waal to the Amsterdam-Rhine Canal, and which are visible from afar due to their size.

A mill or pumping station is a waterwork that brings water from a lower water level to a higher one, and that, via the outlet, connects it to the basin water or the outer water. A mill or pumping station usually drains the water, though sometimes the pumping station is also the site where water is brought into the polder. This is happening more and more often in cases of prolonged drought.
From the fifteenth century onwards, the invention of the windmill with a waterwheel made it possible to pump water out of the lower Netherlands on a large scale. Some are still in use today, and in strong rainfall they still function as reinforcement or backup pumping stations for the usually electrically-driven pumping stations. Most mills are cherished objects of Dutch cultural heritage, and operate during special events, such as the 'mill days' and the country's annual Open Monument Day.
In cases where a lot of water had to be removed, the mills were set up next to each other in a row formation. This is known as parallel drainage, which we see for example at Kinderdijk pp. 160-165. This beautiful ensemble is on the UNESCO World Heritage List, although it is not currently operational, having been replaced by two basin pumping stations at the head of the low basins, and by another basin pumping station between the high basin and the outer waters of the Lek. When there was a large difference in height between the polder water and the basin or outer water, a second mill was sometimes placed next to the existing mill in a staircase formation. This allowed larger differences in height (the number of mills times 1.50 metres) to be bridged, up to a maximum of three metres. Yet this staircase arrangement only really blossomed for the drainage of lakes and ponds, which were up to five metres below the level of the basin. In addition to the lifting mechanism's development from waterwheel to screw-pump pp. 194-195, to vertical or horizontal screw-pump, and later to centrifugal pump, the other main thing that changed was the drive of the pump, from wind to steam and diesel to mainly electric, which is now the most frequently used kind of drive. The introduction of these powerful new techniques greatly increased the capacity and speed at which the water could be pumped upwards. More and more water could be moved with fewer and fewer pumping stations, which were also less visible in the landscape.

The capacity of the pumping stations varies considerably: small pumping stations can pump 50 cubic metres of water per minute, whereas the larger pumping stations, such as the Wouda steam pumping station near Lemmer pp. 72-73, with its iconic chimney, pumps 4,000 cubic metres of water per minute. The electric basin pumping station at IJmuiden is divided into five pumps located next to the Noorder-Zeesluis pp. 46-51, and pumps out 15,600 cubic metres of water per minute.[III] By far the majority of the pumping stations are automatically turned on remotely, a few hours before the expected rainfall begins.

Three types of pumping stations can be distinguished, according to their placement: the ondergemaal (lower-level pump), the polder pumping station, and the basin pumping station. Lower-level pumps are usually located in the middle of the polder, and pump out water from a given water-level zone. These small pumping stations are often little more than green cabinets, and are barely visible, with the exception of the many mills present in the countryside p. 175 and the small pumping stations in Waterland pp. 204-205. Situated along basin networks, polder pumping stations either pump the polder water out to the basin, or pump the polder water directly into the outer water. The Lely pumping station in the Wieringermeer, dating from 1931 pp. 6, 176-179, is an early example of an electric polder pumping station. The pumping station stands monumentally at the point where three polder waterways come together, and pumps the polder water out into the IJsselmeer. This pumping station is no longer in use, and the two other large polder pumping stations in the Wieringermeer have since taken over the work. New pumping stations with improved technology are often placed next to the old ones. In fact, there are three generations of pumping stations pp. 202-203 located side by side in the Ronde. Basin pumping stations are located at the end of the basin system, such as the Schilthuis pumping station in Rotterdam pp. 196-199 at the end of the Rotte, which pumps out the basin water to the outer water.

In addition to pumping stations that are positioned between two watercourses or surfaces, there are also pumping stations that appear to be standing on dry ground. The polder behind them, for example the Noordwaard in the Bieschbosch, can be submerged to serve as a temporary spillway for the river. At a given moment, when the water can be discharged back into the river, these pumping stations pp. 222-227 come into action.

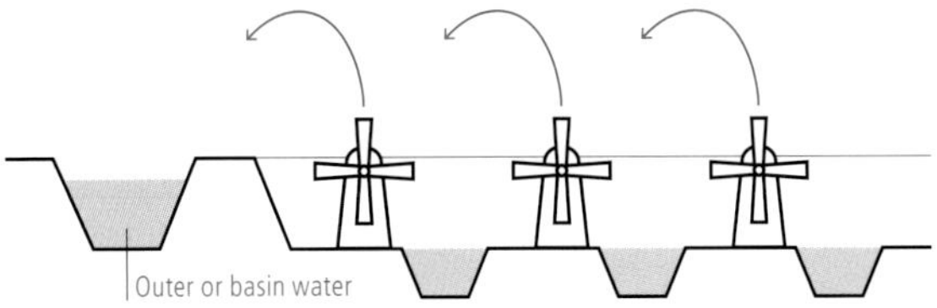

Parallel pumping

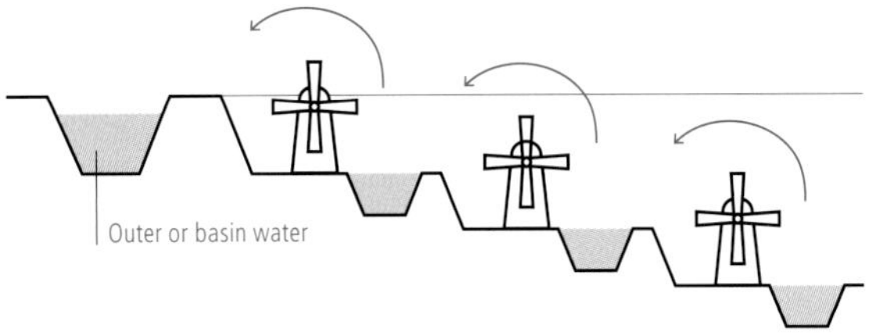

Staircase formation

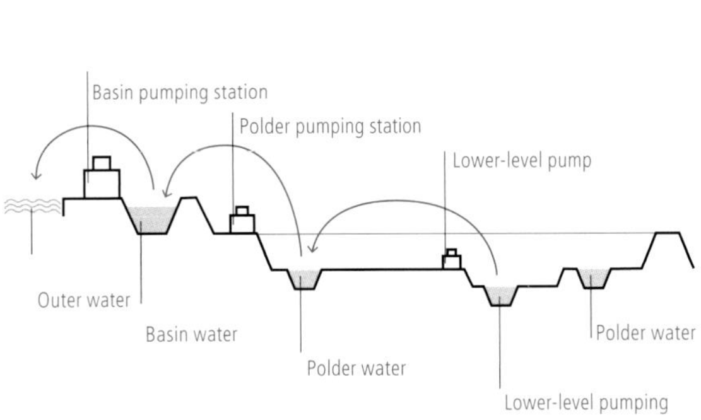

Basin pumping station, polder pumping station, lower-level pump

II. The basin is water that forms part of the basin system, which is a water network that 'mediates' between the lower-lying water in the polders and the higher outer water.
III. This data comes from the Nederlandse Gemalenstichting: www.gemalen.nl.

0
10
20
30
40
50 km

Bleiswijkse Verlaat Overflow, Bleiswijk

The first Bleiswijkse Verlaat was built in the Lange Vaart near the river Rotte during the reclamation of the Bleiswijk polder, which started in 1772. A verlaat (overflow) is a kind of navigation lock, also sometimes called a sas or zijl. Much like the neighbouring Boterdorpse Verlaat, the Bleiswijkse Verlaat is a so-called yoke lock made entirely of wood. A lock cannot be any simpler: the Bleiswijkse Verlaat is not much more than a 30 metre by 3 metre by 3 metre wooden box that rests on 140 spruce piles. Wooden pointed gates close the box on both sides.

The image of the Bleiswijkse Verlaat is defined by twelve high wooden yokes and five slightly lower ones. The yokes, which are intended to keep the lock's gates and floor in place, are less than two metres apart. They give the lock the appearance of a primitive, tunnel-like version of the later Bailey bridges.

The Bleiswijk yoke lock served not only as a navigation lock to allow peat barges and prams to bridge the difference in level between the river Rotte and the Lange Vaart, but also as a flood barrier. By the second half of the twentieth century, the overflow was rarely used as a lock, and it became a ruin.

In 1972, the historic Bleiswijkse Verlaat was restored, or rather reconstructed. Just as some wooden Japanese temples are completely rebuilt after about thirty years or so, so too were all parts of the yoke lock replaced by new ones made of oak.

Thirty years later, it turned out that the oak decayed quickly, which is why the Bleiswijkse Verlaat was reconstructed in 2015. This time the lock, which was declared a national monument in 1974, was built entirely of traditionally grown tropical hardwood.

28 Navigation Lock at the Bleiswijkse Verlaat Overflow, Bleiswijk 1776
Design —

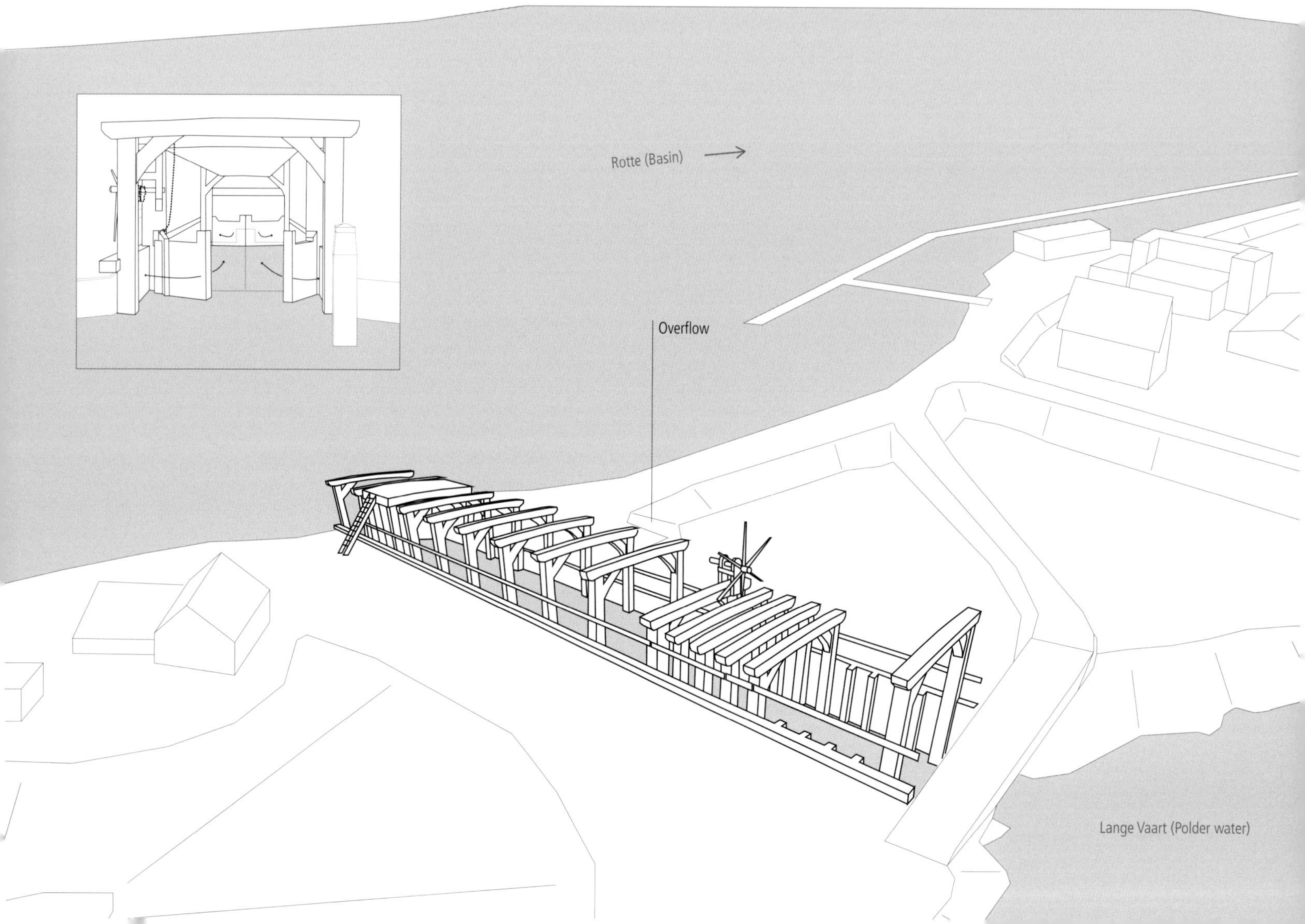

Katerveer Lock, Zwolle

32 Katerveer Lock, Zwolle 1819, 1873
Design —

Oranje Locks, Amsterdam

The building of the Oranjesluizen (Oranje locks) became necessary as a result of the construction of the North Sea Canal in the years after 1865. At the initiative of a number of Amsterdam city administrators, the North Sea Canal was dug for the same reason that the North Holland Canal had been dug half a century earlier: to make the port of Amsterdam more accessible to seagoing vessels. But soon the long Noordhollands Canal (which stretched from Den Helder to Amsterdam) was too small for the ever-larger vessels.

To be able to properly regulate the water level in the North Sea Canal between IJmuiden and Amsterdam, the IJ had to be closed off from the Zuiderzee on the east side. The Oranje locks mark the boundary between the inner and outer IJ, and extend from the Schellingwoude hamlet in the north (where the first stone houses were built for the staff of the Oranje locks) to the Zeeburger Island in the south.

Originally, the Oranje locks consisted of three navigation locks that are now used for recreational ships and small inland vessels. The large lock that was built in 1995 for the large inland vessels was named after Willem-Alexander, who was the Crown Prince at the time. Over the years, the Oranje locks have also been given two fish passages.

Much like the Noorder-Zeesluis on the west side of the North Sea Canal, the Oranje locks form a rather unremarkable whole. The most striking aspect is the control buildings. The one in the Prins Willem-Alexandersluis that was built in 1994, designed by Aletta van Aalst & Partners, a firm specialising in construction work on (water) roads, is clearly related to the older Oranje locks building: the new control building has been given a similar bulge, with sloping cabin windows and the same red colour accents as the old one.

36 Oranje Locks, Amsterdam 1872, 1995
Design J. de Rijke

De Schans Control Lock, Munnekezijl

40 De Schans Control Lock, Munnekezijl 1882
Design ir. Methardus

1882

Bovenlandse Lock, Doeveren

The Bovenlandse lock near Doeveren is part of the 'schans' (entrenchment) that was built around 1500 at the Elshoutse Zeedijk, to defend nearby Heusden. Originally, the lock served as a water barrier, discharge sluice, and flooding lock.

The current lock dates from the middle of the nineteenth century, and probably replaced an older lock that was part of a belt of locks that surrounded the fortified town of Heusden. Located at the crossroads of the Oude Maas river and the Drongelens Canal, the lock has outlets to both of those waterways.

The current Bovenlandse lock, equipped with wooden doors, was built largely in a traditional way out of brick and natural stone.

A nineteenth-century novelty is the iron construction, consisting of two parts. On one side of the concrete bridge deck, which is now almost a century old, is a small lifting tower with gears and chains that raise and lower the barrier gate. On the other side, the lock has two simple rectangular structures, with two wheels and chains, on which two narrow lock gates hang.

After the original steam pumping station in nearby Gansoyen, dating from 1898, was replaced in 1963 by the Netherlands' only pumping station on legs, the Bovenlandse lock was used less and less, and the structure ultimately fell into disrepair. The lock was restored in 2017.

42 Bovenlandse lock, Doeveren 19th century
Design —

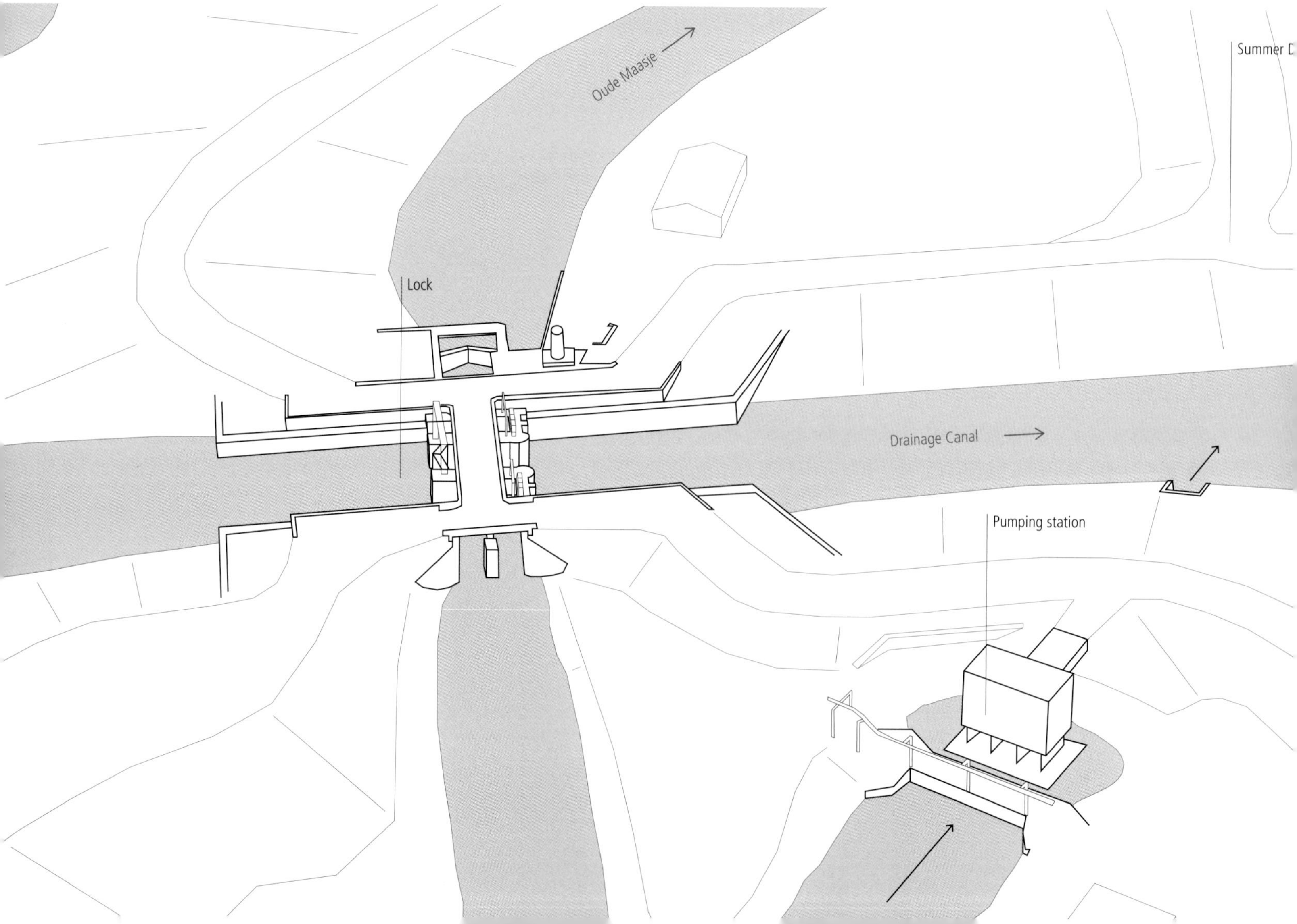

Noorder Lock, IJmuiden

When the Noorder-Zeesluis (Noorder lock) was completed in 1928, it was the largest navigation lock in the world: 400 metres long, 50 metres wide, and 15 metres deep. The Noorder lock had become necessary when the Middensluis (Middle lock), built in 1896 next to the smaller Zuidersluis dating from 1876, became too small, as a result of the ever-larger seagoing vessels that wanted to reach the Amsterdam port via the North Sea Canal. In the 1920s, a new diversion canal was dug in order to facilitate access to the Noordersluis for the ships from the North Sea. As a result of this, the IJmuiden fort became separated from the 'Stelling van Amsterdam' fortifications, of which it had formed a part.

Over the years, the Noorder lock has grown into an extensive complex with simple buildings made of concrete and glass, for example the 1975 pumping station and the small control buildings, with their remarkably thick flat roofs. The Noorder lock is at its most impressive when the giant rolling gates (53.5 metres wide, 7.3 metres deep, and 20 metres high), which roll on invisible rails along the bottom of the lock, close off the chamber.
The Noorder lock has since become too small for the seagoing vessels that are still increasing in size, even today. That is why construction began in 2016 on a lock – 500 metres long, 70 metres wide, and 18 metres deep – between the Middle lock and the Noorder lock. When the lock, designed by ZUS architects, is completed in 2019, it will be larger than the Kieldrecht Lock in the port of Antwerp, which is currently the largest lock in the world.

46 Noorder Lock, IJmuiden 1928
Design Dienst Rijkswaterstaat

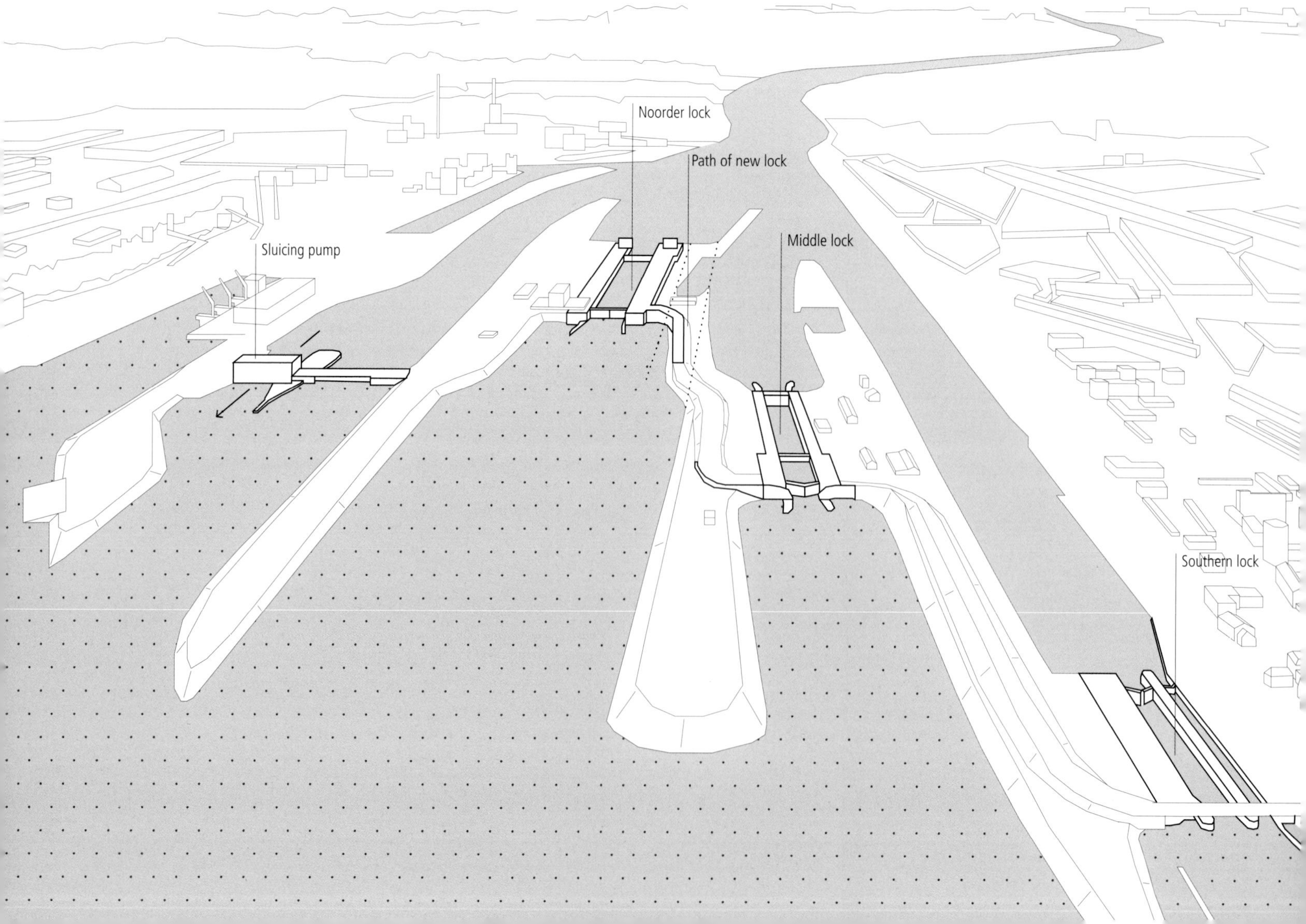

SAFETY FIRST

HERAS

Eefde Lock, Eefde

52 Eefde Lock, Eefde 1933
Design ir. D. Roosenburg

Complex of navigation lock and pumping station; due to the high degree of water deterioration, it is necessary to pump water back into the Twente Canal.

Stevin Locks, Den Oever

Together with the Lorentz locks at Kornwederzand, the Stevin locks at Den Oever, named after the Dutch hydraulic engineer Hendrik Stevin (1614-1668, also the designer of the first plan for the closure of the Zuiderzee in 1667), provide drainage for the IJsselmeer, which is what the Zuiderzee turned into after the construction of the Afsluitdijk in the years 1927-1932. In addition, the Stevin locks serve as a storm surge barrier: in the event of severe storms, the northern slides are lowered to repel the water from the Wadden Sea. To the south of the southern sliders, the lock shafts are equipped with pointed gates pointed towards the Wadden Sea, and which drains water from the IJsselmeer.

In addition to three scouring sluices, the Stevin locks also consist of a navigation canal. The whole complex is part of the Stelling Den Oever, which includes two small islands with casemates to the west and east of the Stevin locks.
With the three scouring sluices of the Stevin locks (and the Lorentz locks), architect Dirk Roosenburg (1887-1962), who has many civil engineering structures to his name, created a monumental entrance to the Afsluitdijk. Each of the three navigation locks consists of six pairs of concrete lifting towers, between which two sets of five steel lock gates can be lifted up and down. Between and alongside the lifting towers, there are two two-lane motorways and, on the Wadden Sea side, a cycle path over the locks.

With its jumpy rectangular elements, vertical ridges, and cantilevered horizontal roof edges, the eclecticist Roosenburg used simple means to give the lift towers the appearance of rugged sculptures. In particular, motorists who drive across the middle lanes between the 36 towers towards Noord-Holland can imagine a miniature version of the Ville Radieuse, or another ideal city that Le Corbusier dreamed of.

56 Stevin Locks, Den Oever 1933
Design ir. D. Roosenburg

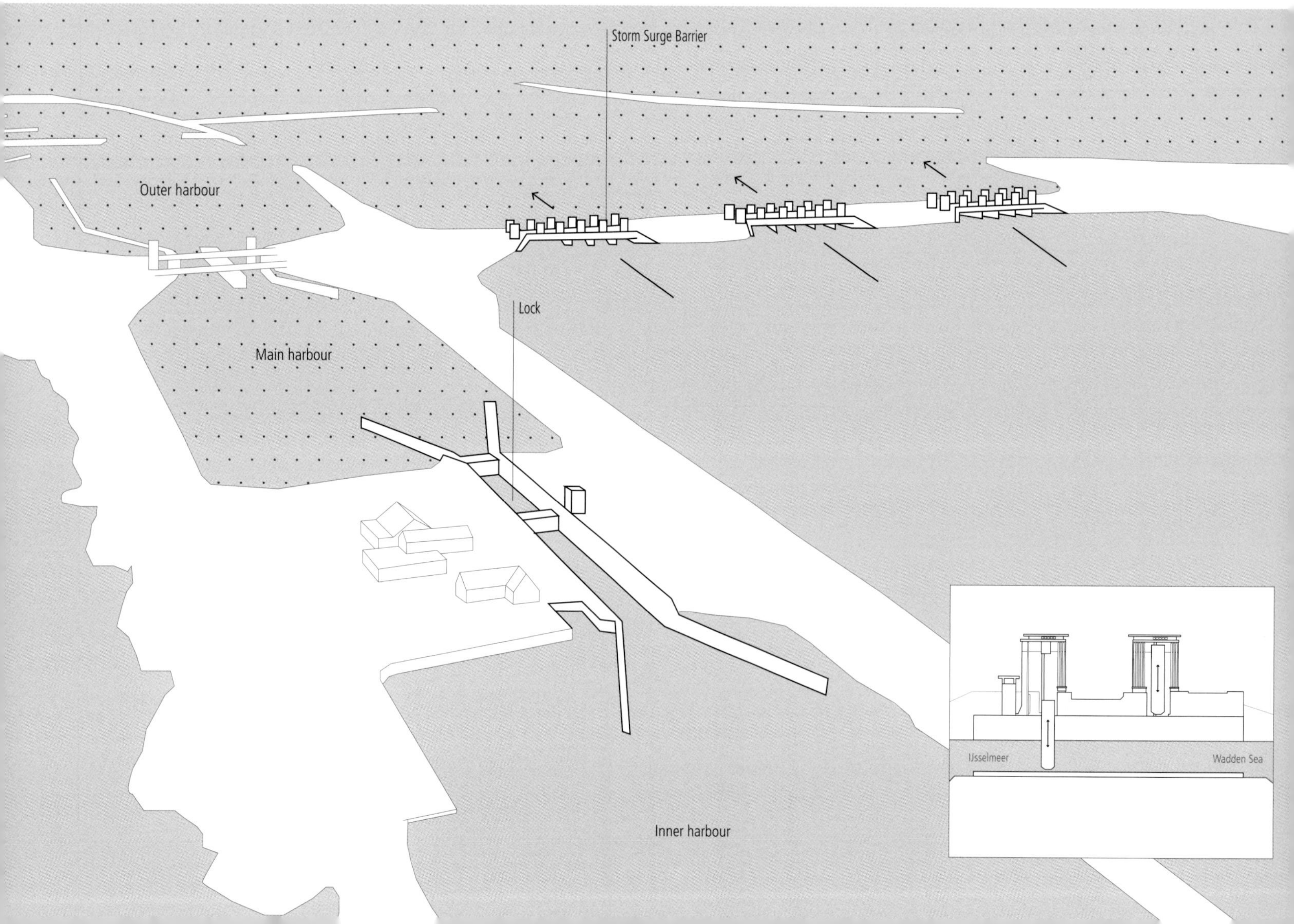

Locks at Born, Maasbracht, Osen, Belfeld, Sambeek

60 Locks at Born, Maasbracht, Osen, Belfeld, Sambeek 1934, 2015
Design – Dienst Rijkswaterstaat

A series of seven locks in the Meuse river, recently renovated and widened as part of VONK (Vervangings-Opgave Natte Kunstwerken; completed in 2012)

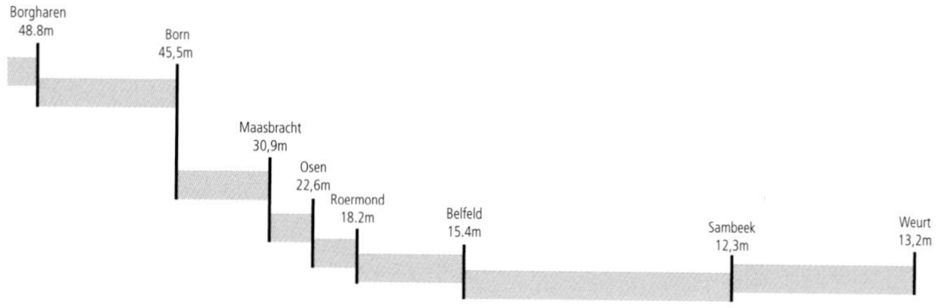

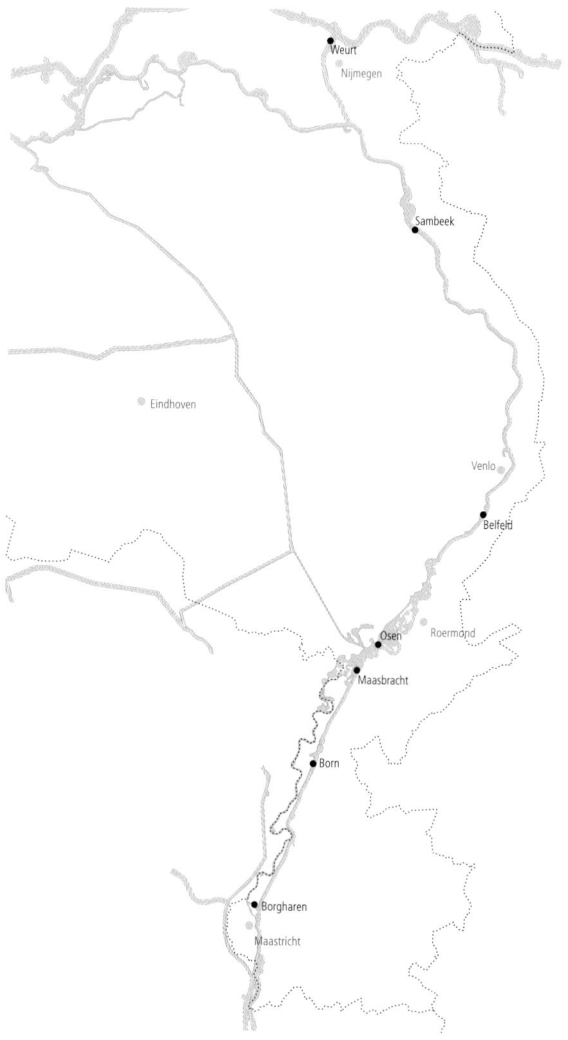

Prinses Beatrix Lock, Vreeswijk

The Prinses Beatrix lock at Nieuwegein bridges the difference in level between the water of the Lek and the Amsterdam-Rhine Canal, which was dug in the 1930s based on a plan drawn up by Anton Mussert, the former Rijkswaterstaat engineer (and later a leader of the National Socialist Movement). This twin lock from 1938 is located in the Lek Canal, which is the connection between the Lek and the Amsterdam-Rhine Canal.

The Beatrix lock, now a national monument, is a genuine engineering work: there was no architect involved. Nevertheless, it seems as if excellent architects worked on it. The two designers of the navigation locks, the Rijkswaterstaat engineers J.P. Josephus Jitta and L.S.P. Scheffer, made it a wonderful example of the Nieuwe Bouwen school. To raise and lower the four steel lift gates, each weighing between 150 and almost 200 tonnes, the two engineers from Rijkswaterstaat's design department created two high concrete bridge buildings, which, as part of the Nieuwe Bouwen project, were given Corbusian 'strip windows' which are almost forty metres long, and which are plastered white from the bottom to the top. The round balconies, which hang like two breasts under the round awnings on the long sides, provide frivolous accents.

With a length of 225 metres and a width of 18 metres, the whole concrete complex is one of the larger locks in the Netherlands. Yet the Beatrix lock, which was renovated and extended in 1997, is still not large enough. A third lock chamber is now under construction. The new lock, which is due to be completed in 2019, will be longer, wider, and deeper than the twin lock from 1938. In order not to detract from the Prinses Beatrix lock, which has been declared a national monument, the third chamber will be built without any tall construction elements.

66 Prinses Beatrix Lock, Vreeswijk 1938
Design J.P. Josephus Jitta and L.S.P. Scheffer

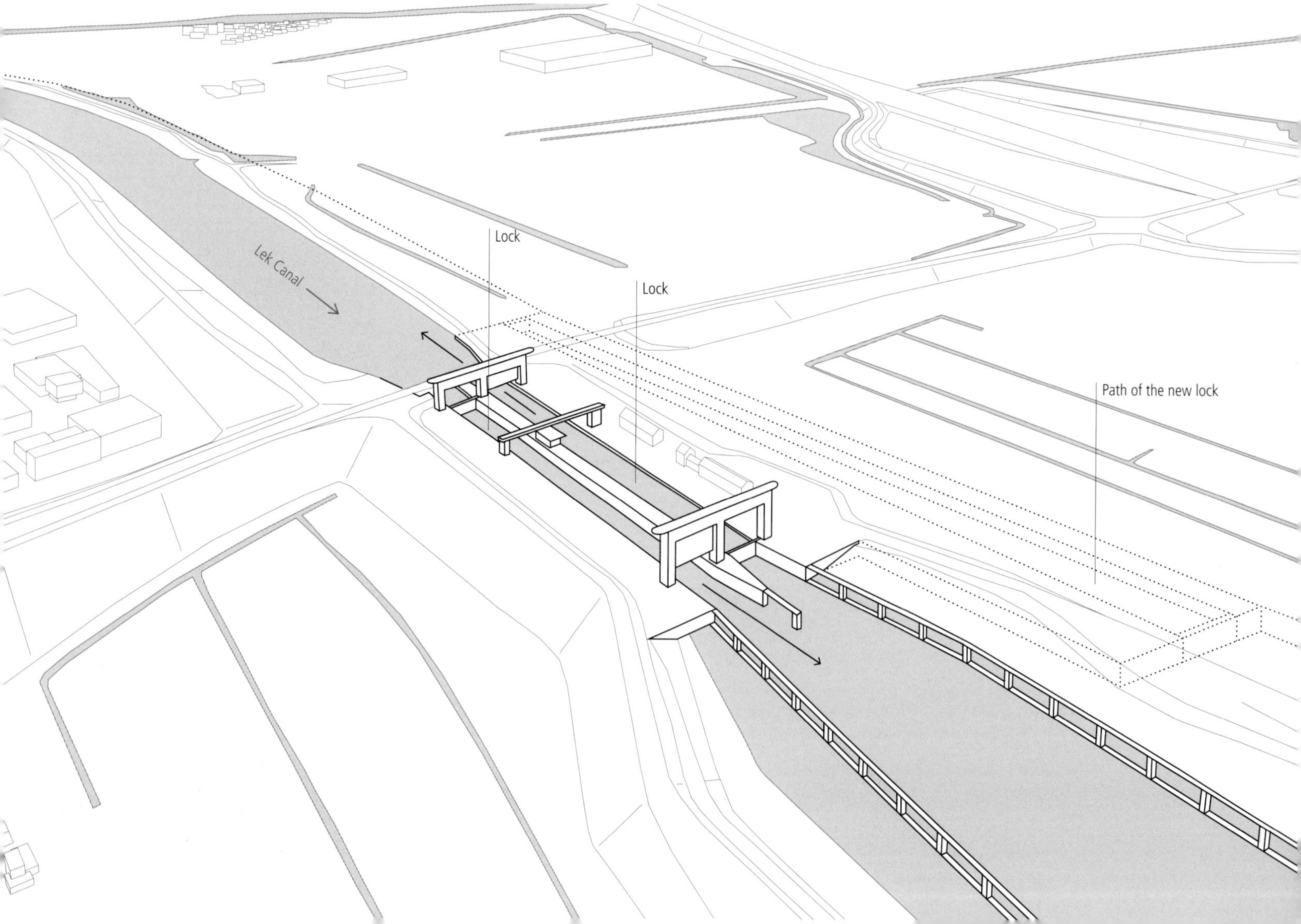

HERMANNA
ZWIJNDRECHT
02317242

heymans
CAT

July 2016

January 2017

May 2017

Prins Bernhard Lock, Tiel

76 Prins Bernhard Lock, Tiel 1952, 1974
Design Dienst Rijkswaterstaat. Lock complex at the connection of the Waal river and the Amsterdam-Rhine Canal.

Prinses Irene and Prinses Marijke Locks, Wijk bij Duurstede, Rijswijk

The Prinses Irene locks at Wijk bij Duurstede (dating from 1952) and the Prinses Marijke locks near Rijswijk are located close together. These locks in the Amsterdam-Rhine Canal have been constructed in front of and behind the intersection of the 72-kilometre-long canal with the Lek. The Irene locks at Wijk bij Duurstede consist of two lock chambers, one of which is 260 metres by 24 metres, and the other of 350 metres by 18 metres; the longer one can be divided into two chambers. The Marijke locks at Rijswijk (Gelderland) have, in addition to the two navigation chambers, a control lock that is only closed when the water in the Lek is at exceptionally high levels.

Both locks are straightforward engineering locks. The Irene locks consist of five lift towers with sliding gates, while at the Marijke locks, only the wide floodgate lock has been equipped with lift towers, between which a gigantic lock gate hangs.

Dirk Roosenburg, the architect of the lifting towers in the Stevin and Lorentz locks in the Afsluitdijk, was involved with the Irene locks as an aesthetic consultant. He remarked on the lift towers and the large dimensions of the locks: 'Finally, a motif on such a large structure that does not disappear into the water, or underground. But this also means that the adaptation to the landscape has also disappeared. The two giant gate buildings in Wijk bij Duurstede are now there, and they dominate the landscape across a large radius. The machine is opened up on the bridge, and is protected from weather influences by metal caps. To be able work in a slightly sheltered way in the upper section, there is a mobile cover on which hangs the hoist, which raises and replaces the machine's parts. It is all honestly visible. An attempt has been made to give these giant gates an honest form, nothing more, but also nothing less than that.'

According to the architectural historian Dorine van Hoogstraten, Roosenburg was too impressed by the new technique of the locks to give them an aesthetic twist. The Irene locks have an 'effective Norwegian look,' she writes in her monograph about Roosenburg.

80 Prinses Irene and Prinses Marijke Locks, Wijk bij Duurstede, Rijswijk 1952
Design Dienst Rijkswaterstaat

rijkswaterstaat
200
EXCELSIOR

88 Ketelhaven Lock and Colijn Pumping Station, Ketelhaven 1957
Design ir. P. Verhave. Combination of a navigation lock and a pumping station. The pumping station is named after a former Dutch Minister of Finance.

PROVINCIE FLEVOLAND
Ketelsluis
Vossemeerdijk 1A
8251 PM Dronten

COLIJN

R.J. Clevering Scouring Sluice, Lauwersoog

The R.J. Clevering scouring sluice (until 2007 known as the Lauwersmeer locks) were built when the Lauwerszee was dammed at the end of the 1960s, and transformed into the Lauwersmeer. Two of the three locks, each with four discharge channels, are located in Friesland, and the third is in Groningen. The water of the Lauwersmeer, in which the Friesian basin and the so-called Electra basin (which is the basin of a number of Groningen polders) end up, is discharged when the water level in the Wadden Sea is lower than it is in the Lauwersmeer. This means that it is not possible to discharge water in the event of a heavy north-western storm, which causes the water in the Wadden Sea to collide against the dam of Lauwersmeer. That is why it was suggested in 2012 to replace the R.J. Clevering scouring sluice with an electric pumping station. But due to the high costs and the expected low usage, the pumping station was eventually cancelled.

The Clevering locks form a strange kind of 'architecture parlante'. The blue panels between the narrow concrete sections of the lifting towers resemble lock gates in their highest position, and clearly say: this is a lock. At the same time, the three tapering structures, reminiscent of mastabas, resemble pumping stations due to their closed nature. The locks also show that waterworks are subject to fashion. Much like the influences of the Nieuwe Bouwen in Dirk Roosenburg's Stevin locks indicate that they stem from the 1930s, these blue panels reveal that the Clevering locks were built at the end of the 1960s, when system construction was at its height in Dutch architecture.

92 R.J. Clevering Scouring Sluice, Lauwersoog 1970
Design Dienst Rijkswaterstaat

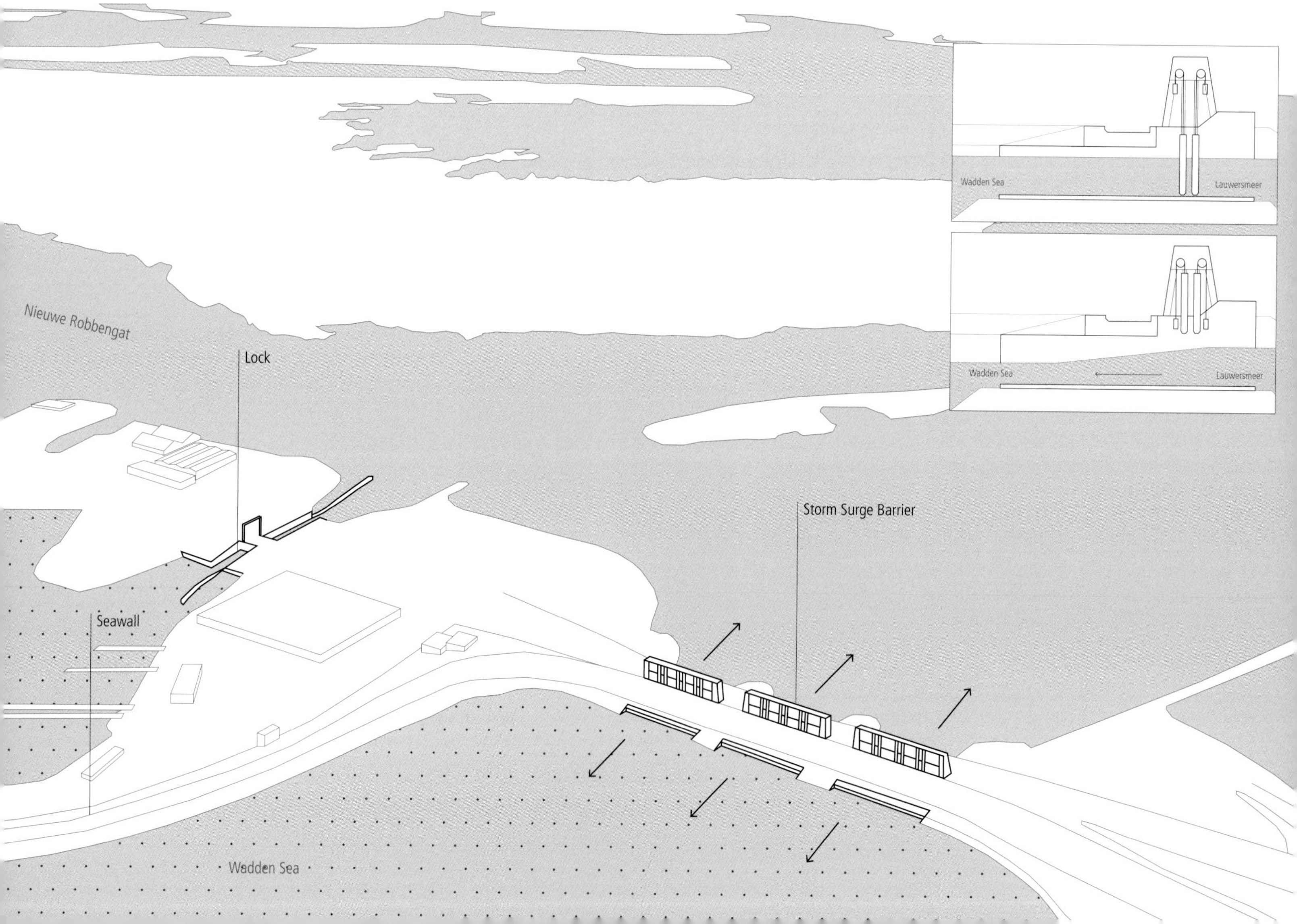

Gemeente
Dongeradeel
FAIRTRADE
GEMEENTE
60

Krammer Locks and Volkerak Locks, Philipsdam, Hellegatsplein

The Krammer locks are located in the Philipsdam in Zeeland, a part of the Delta Works that closes off the Krammer and Volkerak from the Oosterschelde. In 1994, two locks for pleasure craft, measuring 75 by 9 metres, were added to the original twin lock with rolling gates dating from 1983. With chambers of 280 metres long and 24 metres wide, the Krammer locks are among the larger locks in the Netherlands.

The Krammer locks, with outer ports of 1.3 kilometres long and water basins on both sides of the Philipsdam, form a wide, dispersed complex of dams and peninsulas. The largest island houses a control building as well as five pump houses that are reminiscent of mastabas from ancient Egypt due to their strange, slightly upward-sloping shapes. A bridge with a passage height of 18.5 metres has been built over the lock for traffic on the dam road. A lookout tower with a large rectangular platform was built along the road in 1994.

A complicating factor in the construction of the Krammer locks was that there needed to be a limit to how much saltwater from the Oosterschelde would flow into the freshwater Volkerak. The locks are therefore equipped with a system that takes advantage of the fact that saltwater is heavier than freshwater. Depending on the direction of navigation of the ships, the underlying saltwater is sucked away and freshwater is added, or vice versa.

The Volkerak locks, which were built in 1967 at Willemstad, a fortified seventeenth century town, allow the brackish water from the Volkerak to be separated from the water coming from the Hollandsch Diep by means of a 'bubble screen.' Air bubbles, blown into the bottom of a lock by a compressor across the full width of the chamber, form a curtain. Because the rising bubbles on both sides of the curtain cause vertical currents, a screen is created between the different types of water from the Hollandsch Diep and the Volkerak.

When high water conditions occurs in the Hollandsch Diep, the locks are also used to discharge the excess water into the Volkerak-Zoommeer. To this end, the complex also includes four scouring sluices and, after the 1977 expansion, an inlet lock as well.

 Krammer Locks and Volkerak Locks, Philipsdam, Hellegatsplein 1983-1987, 1967-1977
Design Deltadienst

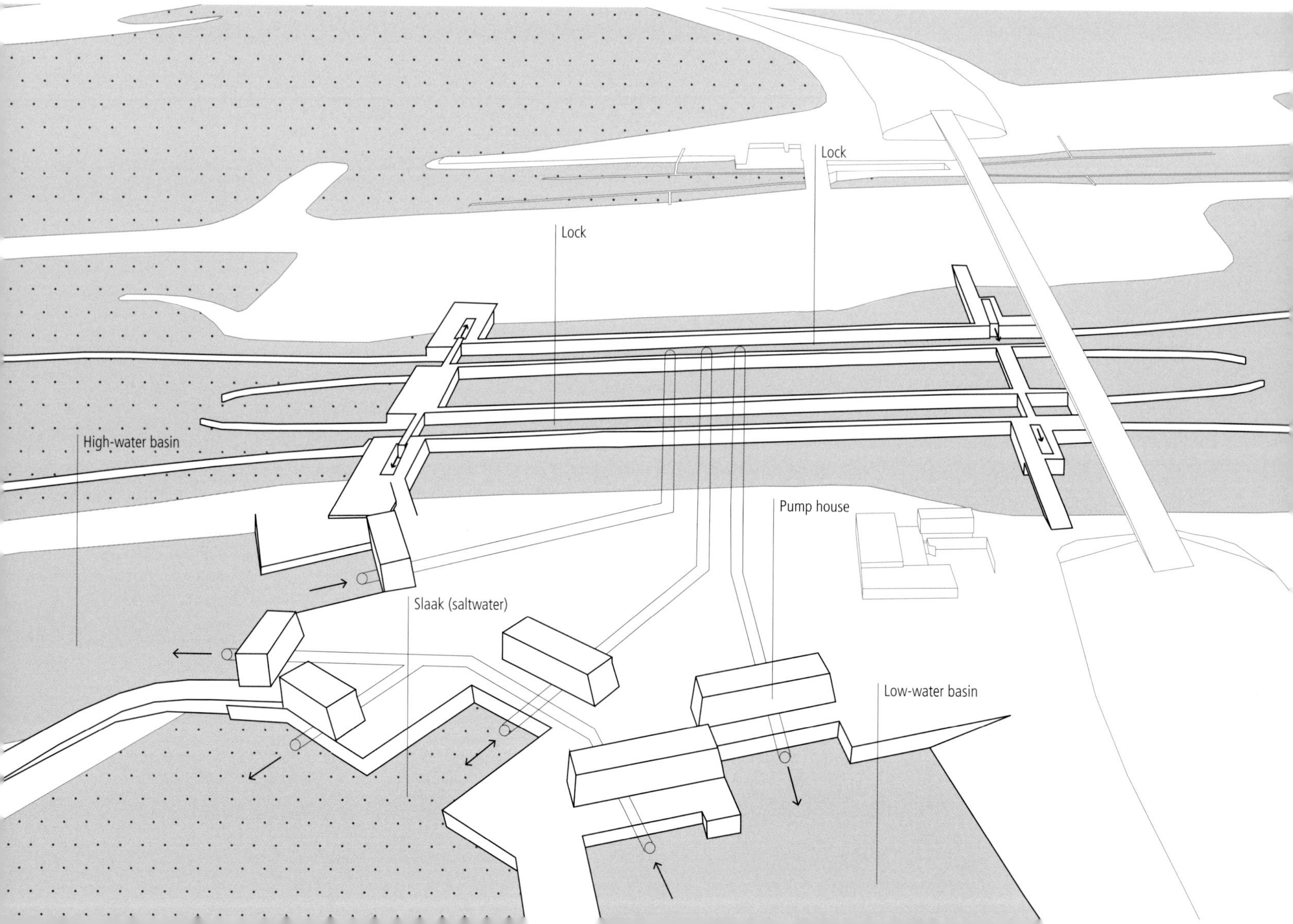

The Volkerak locks, built from 1957 to 1967 as part of the Volkerakdam and extended in 1977, include three locks for commercial shipping and one for pleasure craft. At a length of about 330 metres and a width of 24 metres, and with 48 lock gates, the Volkerak locks are the largest inland navigation locks in the world. A bascule bridge over the locks, part of the A29 motorway, and a concrete viewing platform, also form part of the lock complex.

Since its renovation in 2009, the control building has become the most prominent part of the Volkerak locks. Based on a design by DP 6, a high-contrast, deconstructivist control building was built in that year. Much like the windows of the original straightforward control building (and its ship's cabins), the three glass facades of the building, resting on walls of brick basalt blocks, lean forward. On the other hand, the southern facade of zinc is, apart from a door and narrow triangular windows, completely closed off, and forms a whole with the slightly sloping roof that ends in a wide awning above the glass facades. Because the control building is small in relation to the large lock, the architects of DP 6 have scenically embedded the building in the intersection of waterways and roads that the Volkerak locks are, by including paths and ramparts.

MERCATOR
ANTWERPEN
6004185

SINBAD

Termunterzijl Lock and Rozema Pumping Station, Termunterzijl

102 Termunterzijl Lock and Rozema Pumping Station, Termunterzijl 1725, 2000
Design ir. Cremer, Jongsma and Dijkhuis Architectenbureau

CREMER

Bath Scouring Sluice, Bath

The first thought that emerges from seeing the Bath scouring channel: Why does the largely linear scouring channel run parallel to the Rhine-Scheldt Canal at such a short distance? The answer is simple: the Bath scouring channel, which has a bend to the west precisely at the point where the Rhine-Scheldt connection bends off to the east, serves a different purpose than an ordinary canal does. The scouring channel, which is 8.4 kilometres long, 140 metres wide and 7 metres deep, was constructed from 1980-1987 to allow the water that is used to flush the more northerly dammed waters, such as the Volkerak and Zoommeer, to be discharged into the Westerschelde.

The Bath scouring channel ends at the Bath scouring sluice. Much like a bunker, it is largely hidden in the dike. The lock consists of six concrete shafts, which allow the water to flow into the Westerschelde without the aid of pumps.

106 Bath Scouring Sluice, Bath 1987
Design Deltadienst

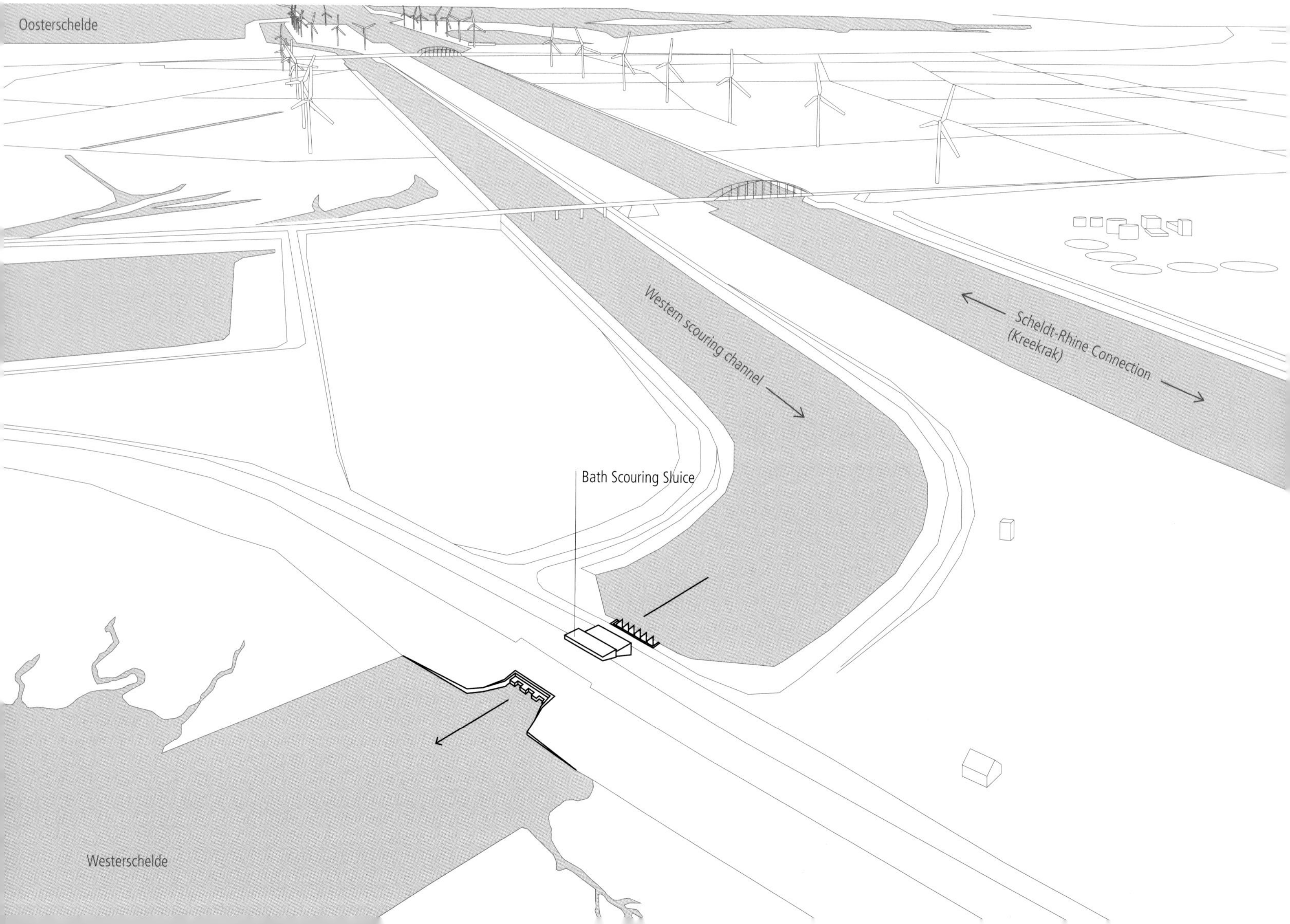

Lock and Weir, Grave

Combinations of bridges, weirs, and locks are common in the Netherlands. But what makes the lock and weir in the river Meuse at Grave unique is that the bridge and weir have been designed in 1929 as a unified whole. The complex, designed by engineer C.F. Egelie of Rijkswaterstaat and expanded in 2006 to include a fish ladder, is also a national monument.

The traffic bridge at Grave, built in 2004, runs along the N324 from Den Bosch to Nijmegen, and consists of nine simple arch bridges of 50 or 60 metres in length. Weirs are hung beneath two of them.

The weir at Grave is the sixth of seven weirs that were built in the 1920s and 1930s to tame the erratic rain-fed river Meuse, from Limburg to Lith. The weir consists of two sets of ten steel yokes with three sliders each, which can be positioned as needed depending on the water levels. In particularly low water conditions, all 60 sliders are placed on the yokes. When the discharge is more than 1000 cubic metres per second, all the yokes hang in a horizontal position under the two sections of the bridge, so that ships can sail underneath. When the discharge is less than 1000 cubic metres per second, the ships have to use the 142-metre by 16-metre lock that was built in 1974 next to the original lock of the complex near Grave, which has since fallen into disuse.

The weir at Grave sets the water level in the important Meuse-Waal Canal. The importance of this weir was shown when a ship sailed through the closed weir in dense fog on 29 December, 2016. As a result, the Meuse between Sambeek and Grave emptied out, and the Meuse-Waal Canal became unnavigable.

110 Lock and Weir, Grave 1929
Design ir. C.F. Egelie (Rijkswaterstaat)

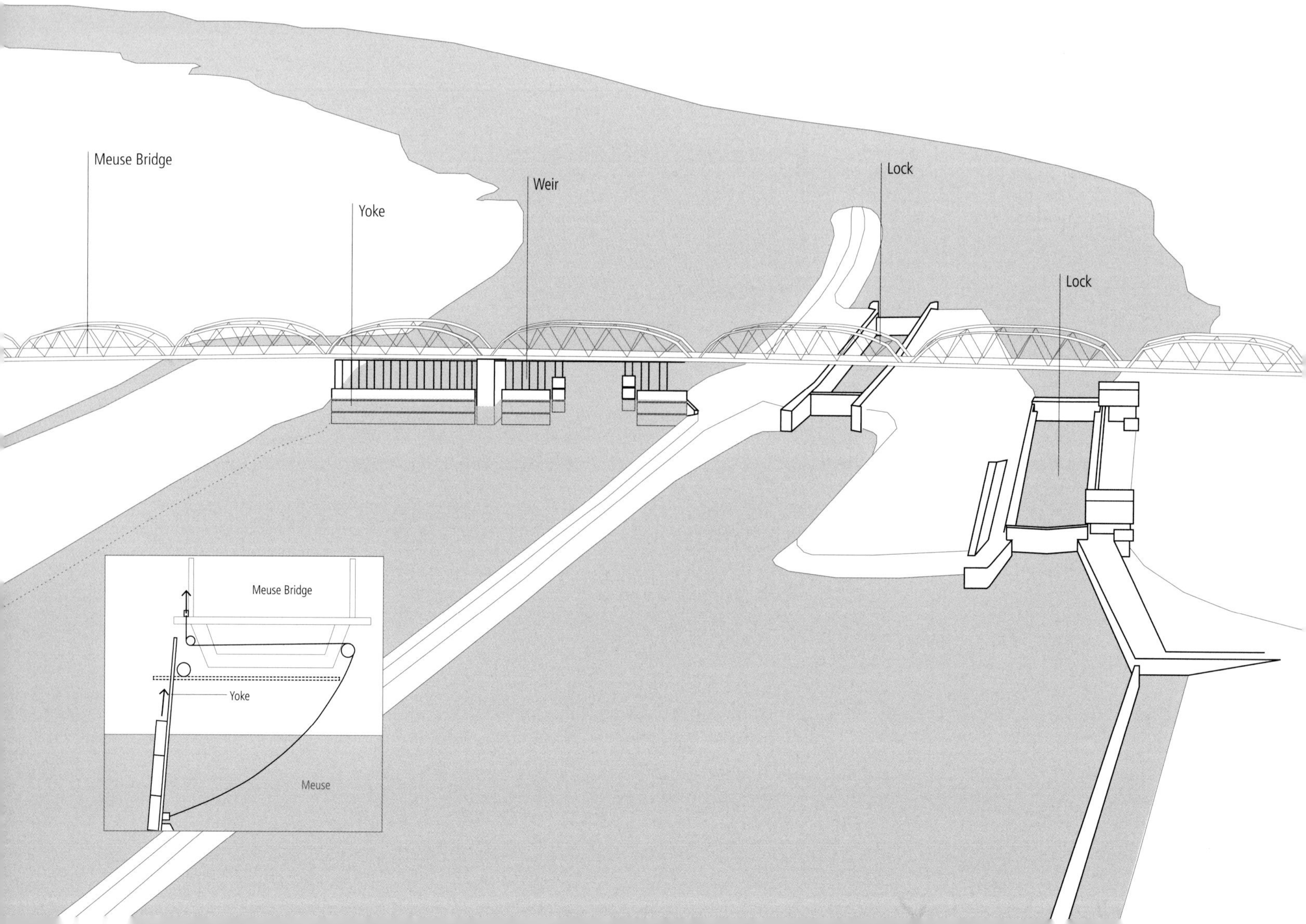

Algera Storm Surge Barrier, Krimpen aan den IJssel

The Stormvloedkering Hollandse IJssel (Hollandse IJssel storm surge barrier), also known as the Algerakering, in the Hollandse IJssel at Krimpen aan den IJssel, was the first civil engineering structure to be built within the framework of the Delta Works. In January of 1954, less than a year after the great flood of 1953, construction began on the Stormvloedkering. Four years later, the barrier was completed, and was named after Jacob Algera, the former minister of Verkeer en Waterstaat (now known as the Ministry Transport, Public Works and Water Management).

The Algerakering is a double barrier. With two sets of two 45-metre-high lifting towers relatively close to each other, it forms an impressive port to the Dutch delta region. Between the lifting towers are two sliders (a 'primary' one and, for safety's sake, a reserve) that are 82 metres wide and 11.5 metres high. They can be lowered into the river in cases of extremely high water levels. As is the case with almost all other weirs, a lock has also been constructed at the Algerakering, and the complex also has a truss bridge.

The towers and weir gates of the Algerakering form a less dramatic composition than do the weirs of Driel, Hagestein and Amerongen. But with the overhanging facades of the upper parts of the towers, the sloping stairs, and the convex truss on the rear of the doors (as if a bridge that had been rotated a quarter turn were hanging between the towers), the designers H. G. Kroon and J. A. G. van der Steur did manage to make it a dynamic whole.

The weir is at its most beautiful at night. When the gates are raised, the towers are illuminated in blue, and in red when the gates are lowered.

The construction of the Maeslantkering in the more westerly Nieuwe Waterkering in the 1990s did not render the Algerakering superfluous. The Maeslantkering is only operated during extremely dangerous conditions. In the years 2013-2017, the Stomvloedkering Hollandse IJssel was operated five times due to storms and/or spring tide.

114 Algera Storm Surge Barrier, Krimpen aan den IJssel 1958
Design H.G. Kroon and J.A.G. van der Steur

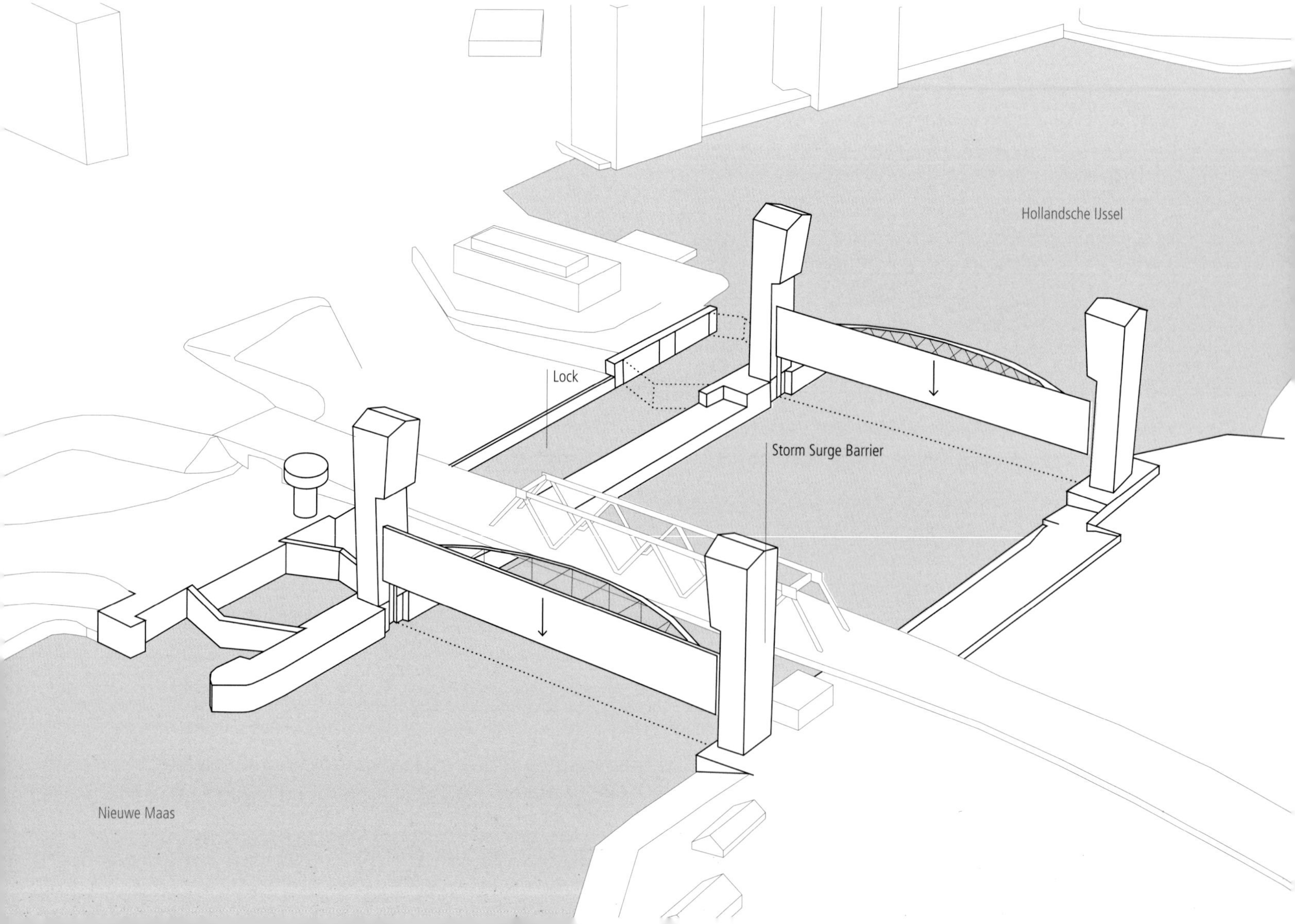

Weirs at Driel, Amerongen and Hagestein

Together with the almost identical weirs near Hagestein and Amerongen, the weir at Driel forms a complex in the Lower Rhine and the Lek that was built starting in 1958 to regulate the (fresh) water in the river area and the IJsselmeer. Of these three weirs, which were built in the years 1958-1970, the first and most eastern one, at Driel, is called 'the faucet of the Netherlands.'

The three weirs, which are equipped with fish ladders as well as locks are that between 225 and 260 metres long, divide the Rhine water between the Lower Rhine, the Lek, and the IJssel. At normal and low water levels, the weirs are closed; at high tide, the colossal convex weir gates are lifted out of the water in order to ensure that the water from the Rhine is discharged to the sea as quickly as possible.

All three complexes were built based on a design by the architect Wouter Hamdorff (1890-1965). Most of Hamdorff's country houses, water towers, and museums are built in a style that is halfway between the Amsterdam School and the pre-war style of W.M. Dudok. With the weirs in the Lower Rhine and Lek, Hamdorff also refers to the Nieuwe Zakelijkheid, the style that fits so well with civil engineering structures. Especially when the convex gates have been lifted out of the water, the lifting towers, with their sloping and rounded shapes, form dynamic odes to the ever-flowing waters of the Rhine and Lek.

In the years 2016-2019, all three weirs will be renovated and expanded with new control buildings, based on a design by Powerhouse Company. Using buildings with round shapes and facades with crystalline reliefs, Nanne de Ru of Powerhouse Company has provided architecture that integrates well with Hamdorff's masterpiece.

120 Weirs at Driel, Amerongen and Hagestein 1960, 1965, 1970
Design W. Hamdorff

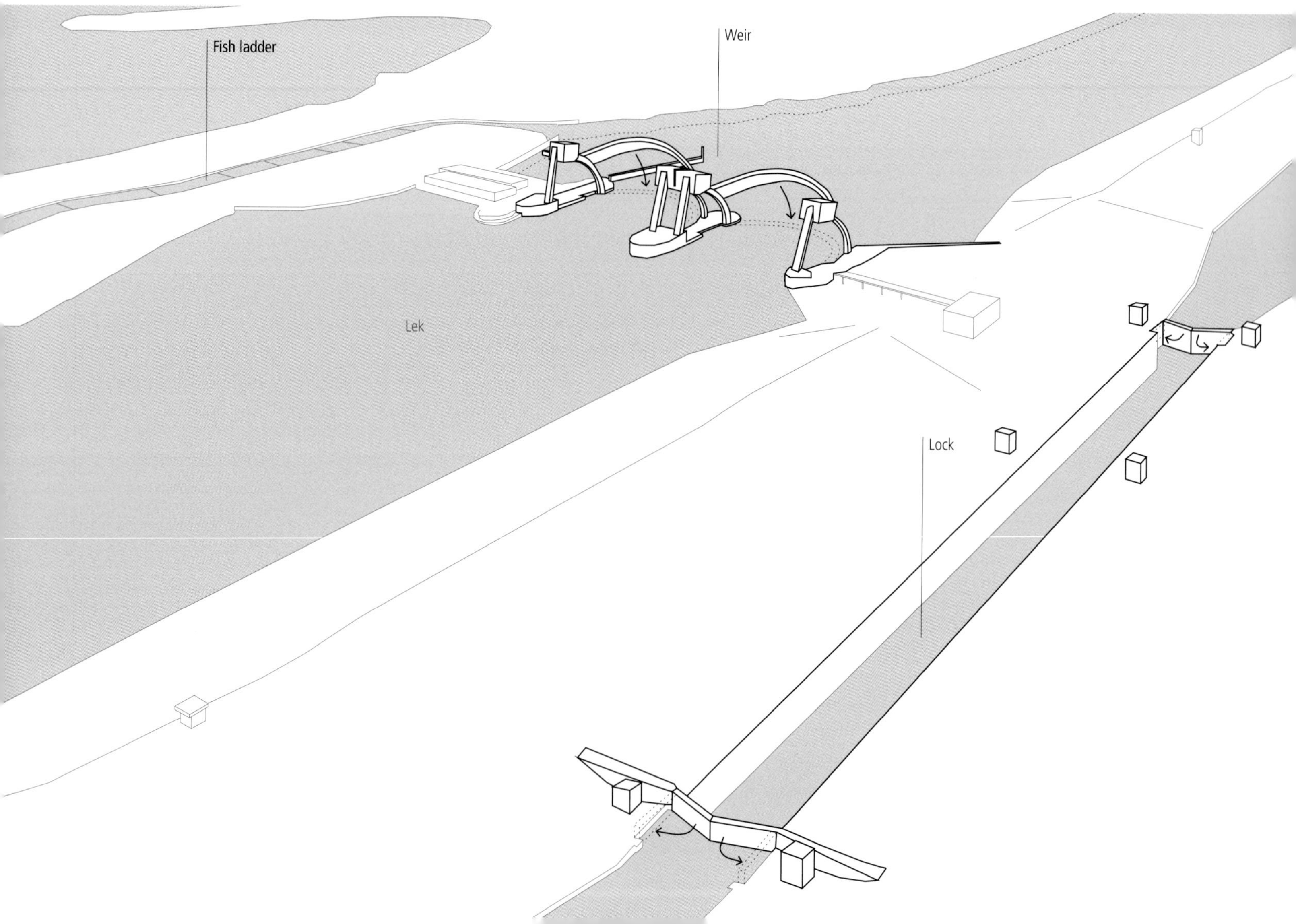

Storm Surge Barrier, Oosterschelde

The Oosterschelde Stormvloedkering (storm surge barrier) offers the joys of repetition. Driving along the nine-kilometre-long road of the storm surge barrier from Schouwen to Noord-Beveland, one passes 65 almost identical concrete, sand-filled pillars, all of which are fitted with pipes. Between the pillars, and divided over the three sections (the longest of which is adjacent to Noord-Beveland), 62 sliding gates that are each 42 metres long and six to twelve metres high are suspended. Only when the water of the Oosterschelde threatens to rise higher than three metres above NAP are those gates slid downwards to block the rolling water.

The Stormvloedkering is the result of the Dutch back-and-forth style of negotiations. At the end of the 1960s, construction began on a dam that would completely close off the Oosterschelde. When that dam, with the artificial islands of Roggenplaat (1969), Neeltje Jans (1970) and Noordland (1971) was already two-thirds finished, the work was halted in 1974 because of mounting protests from nature organisations, among others. In order to limit the impact on nature that a complete closure in the Oosterschelde would cause, the Den Uyl cabinet decided in 1976 to complete the dam with barrier defences that are almost always in the open position. This would maintain three-quarters of the tidal changes in the Oosterschelde behind the dam, and would also allow the water to remain saltwater.

In 1986, Queen Beatrix opened the storm surge barrier, which is by far the largest and most expensive of all the Delta Works, with the words: 'The storm surge barrier is finished. The Delta Works are completed. Zeeland is safe.' In fact, the Delta Works would not actually be completed until the Maeslantkering was finished in 1997.

A control and exhibition building designed by Wim Quist was built on the Neeltje Jans production platform. As a counterpoint to this extremely sober rectangular 1985 building, named after Ir. J.W. Tops, the undulating Water Pavilion appeared in 1997 after Neeltje Jans was gradually transformed from a production platform into a nature reserve and amusement park. It consists of two whimsical, fully curved blobs designed by NOX Architects and Oosterhuis Associates.

126 Storm Surge Barrier, Oosterschelde 1986
Design Deltadienst

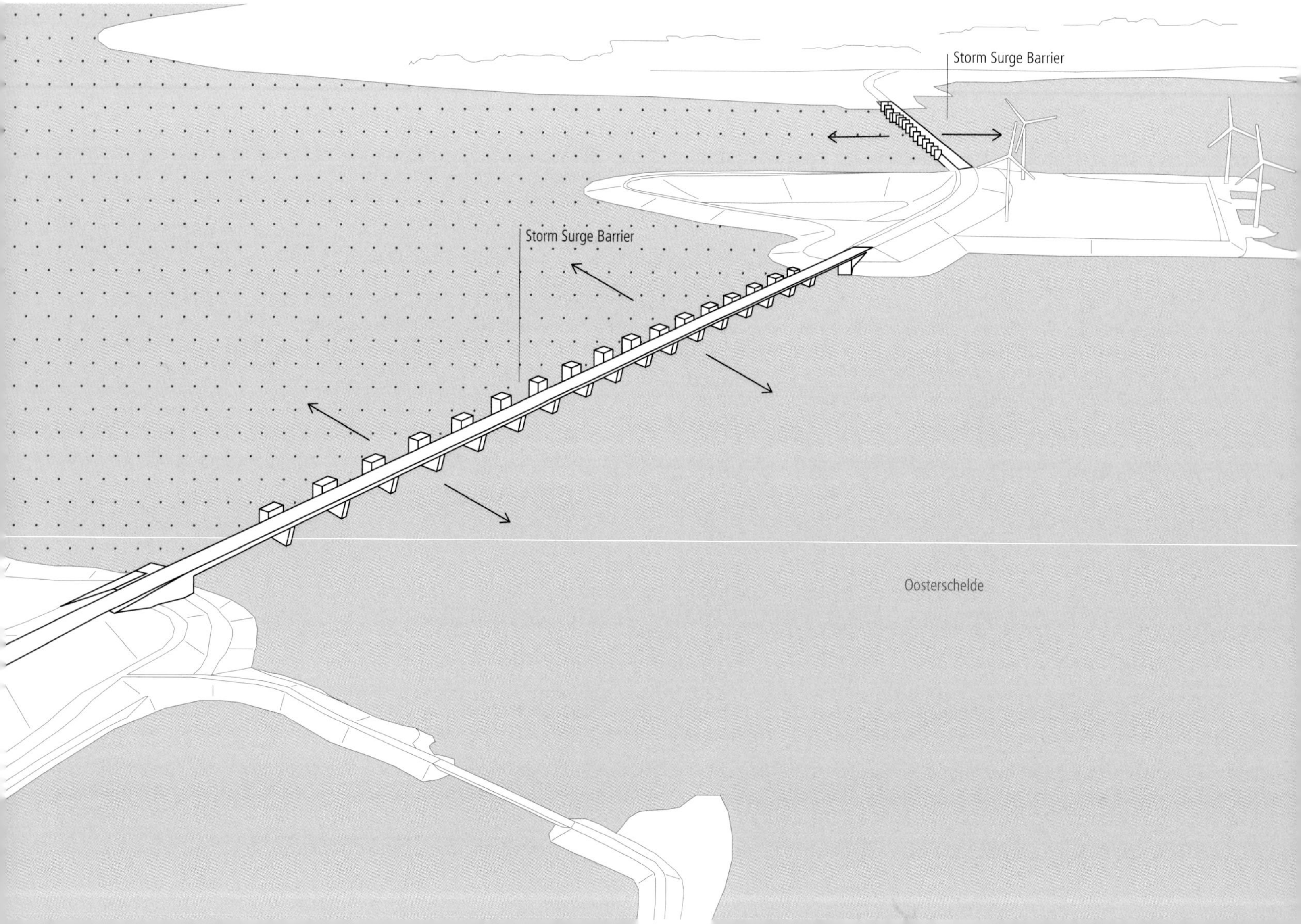

S5

Maeslant Storm Surge Barrier, Hoek van Holland

The figures make it sound like one of the wonders of the world: the 'Maeslantkering' in the Nieuwe Waterweg is almost as long as the Eiffel Tower (317 metres) is tall. The two curved 'repelling walls' are each 22 metres high and 210 metres long; the v-shaped truss arms, which move the gates when there is an approaching flood of at least three metres above NAP, are 237 metres long. The lower two of the three pipes of the truss arms are as tall as a person (1 metre 80), and the spherical hinges required to turn the arms have a diameter of 10 metres, with each weighing 680 tonnes. Invisible is the barrier that is built into the 17-metre-deep Nieuwe Waterweg, consisting of 64 concrete blocks each weighing 630 tonnes. The hollow floodgates come to rest on them after they have been flooded with water, and will thus be able to stop a flood of up to five metres above NAP

The Maeslantkering is not a pure engineering work by Bouwkombinatie Maeslant Kering, which in 1991 won the competition for a flood defence barrier at Hoek van Holland.

In order to make this movable civil engineering structure even more beautiful than it already was, Wim Quist, an experienced architect in the field of these waterworks, was involved in the design. Not only were the control buildings of the barrier designed by Quist's office, but that office also gave advice on the 'visual aspects' of the barrier, such as the colour of the truss arms.

The combination of engineering-art and architecture has in this case resulted in sublime beauty. Like an awe-inspiring version of a Theo Jansen beach animal, the two truss arms wait patiently for the signal to set in motion the colossal gates, which are ensconced in concrete dry-docks. The arms have been receiving that signal once per year since the barrier was completed, during the 'operating closure' that takes place annually shortly before the storm season begins (1 October - 1 April). It is at those moments that the Maeslantkering, like a male peacock displaying its tail, is at its most beautiful.

132 Maeslant Storm Surge Barrier, Hoek van Holland 1997
Design Bouwkombinatie Maeslant Kering

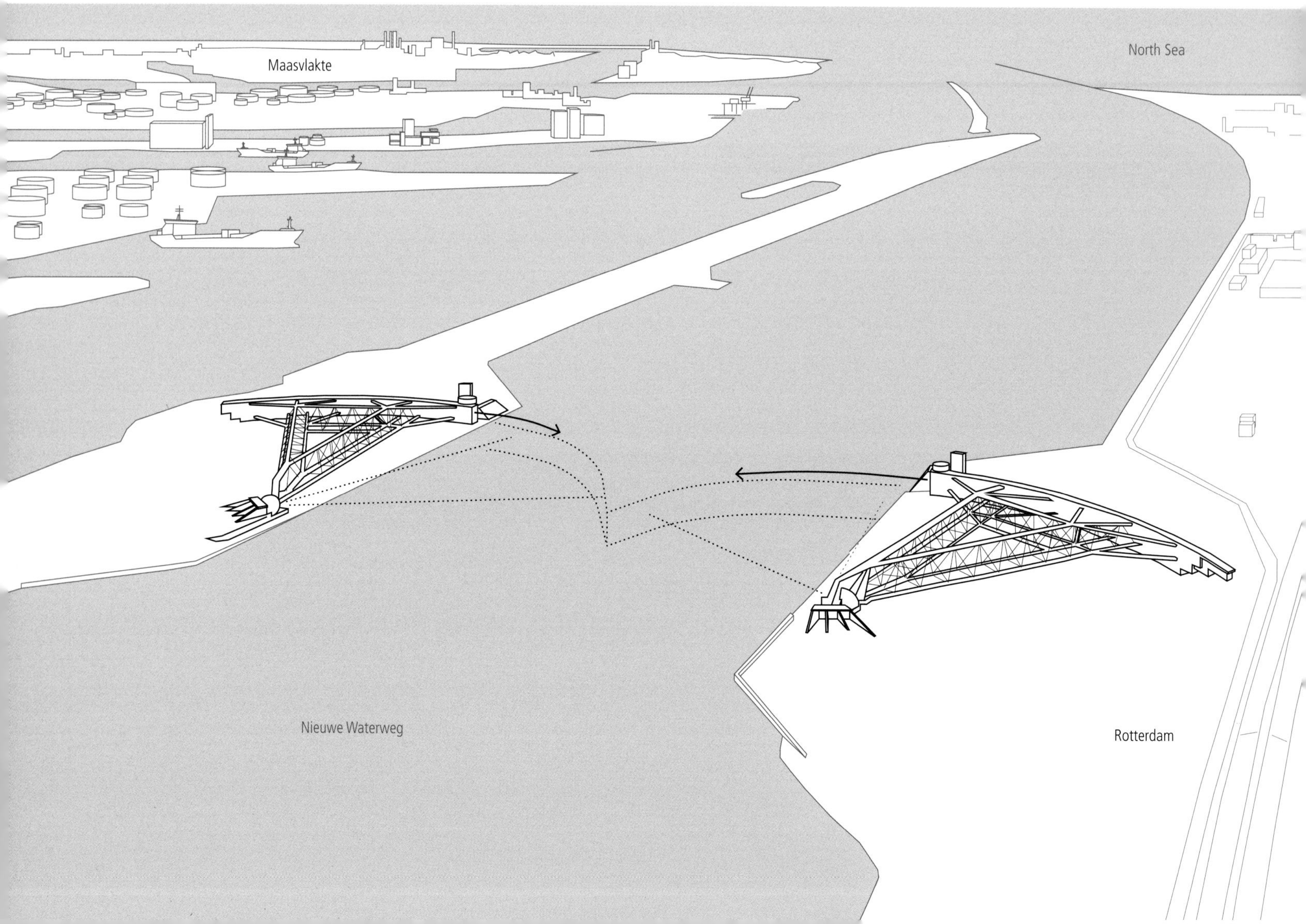

De Kromme Nol Control Lock and Bridge, Wijk en Aalburg

The Kromme Nol control lock is the result of the exceptionally high water levels in the (large) rivers in January 1995, when parts of Limburg were flooded, and 250,000 people has to be evacuated in the river area due to the looming floods. This led to the Deltaplan Grote Rivieren (Delta Plan for the Big Rivers), which provided for the heightening of the dikes along the Dutch rivers. Later, the 'Room for the River' Key Planning Decision (Planologische Kernbeslissing Ruimte voor de Rivier) took effect. The emph asis is on interventions such as increasing the size of the flood plains, and the construction of retention basins.

The Kromme Nol control lock was built in the Heusdens Canal, which connects the Afgedamde Maas to the Bergense Maas. The barrier and the associated bridge were built to a design by Quist Wintermans Architects on the site of a Bailey bridge, which had been built in 1945 after the first bridge over the Heusdens Canal (dating from 1896) was bombed in 1944.

Quist Wintermans Architects have made something like mini-skyscrapers, complete with needles, out of the two lifting towers of the Kromme Nol floodgate, between which a straight black sliding gate hangs. With their round sides, replete with windows that provide light in the staircase, the lifting towers somewhat resemble the Flatiron Building in New York, dating from 1902. The ornamental horizontal grooves on the sides of the gates, which enhance the skyscraper effect, are intended as references to the mechanism by which the gates are slid up and down during periods of high water.

The concrete bridge, a slightly curved deck that rests on two pillars, has been kept as simple as possible in order to avoid having to compete with the monumental barrier. There is a monument at the bridge, which consists of a section of the Bailey bridge, commemorating both the older bridge and the Second World War.

138 De Kromme Nol Control Lock and Bridge, Wijk en Aalburg 2001
Design Quist Wintermans Architects

Inflatable Weir, Ramspol

It cannot be a coincidence that Zwarts and Jansma Architects have given the two control buildings of the Balgstuw at Ramspol a crustaceous appearance. This inflatable weir, which was built in the early twenty-first century between the Ketelmeer and the Zwarte Meer, is actually a kind of animal. It is usually asleep, but it opens up when a north-western storm pushes the water from the Ketelmeer, which washes against the Black Lake, to a height of half a metre above the Nieuw Amsterdams Peil level. The pumps are then switched on from the small control buildings, which are covered with five conical stainless steel shells. They allow three 80 metre-long bellows made of rubber fabric, which lie at the bottom of the Zwarte Meer, to fill up with 21 million litres of air and water, until they (together with the fixed parts of the weir) form a 10-metre-high surge barrier.

The Ramspol inflatable weir is the largest one in the world to also serve as a surge barrier. Rijkswaterstaat chose to build this 100 million euro weir because the alternative, namely heightening the Overijssel dikes along a length of 115 kilometres, was much more expensive. It is expected that the inflatable weir will need to be activated on average once per year. This average has not yet been borne out: in 2015, the weir-animal was awakened from its sleep three times, but this had previously only happened twice, in 2007 and in 2012.

142 Inflatable Weir, Ramspol 2002
Design Zwarts & Jansma Architects

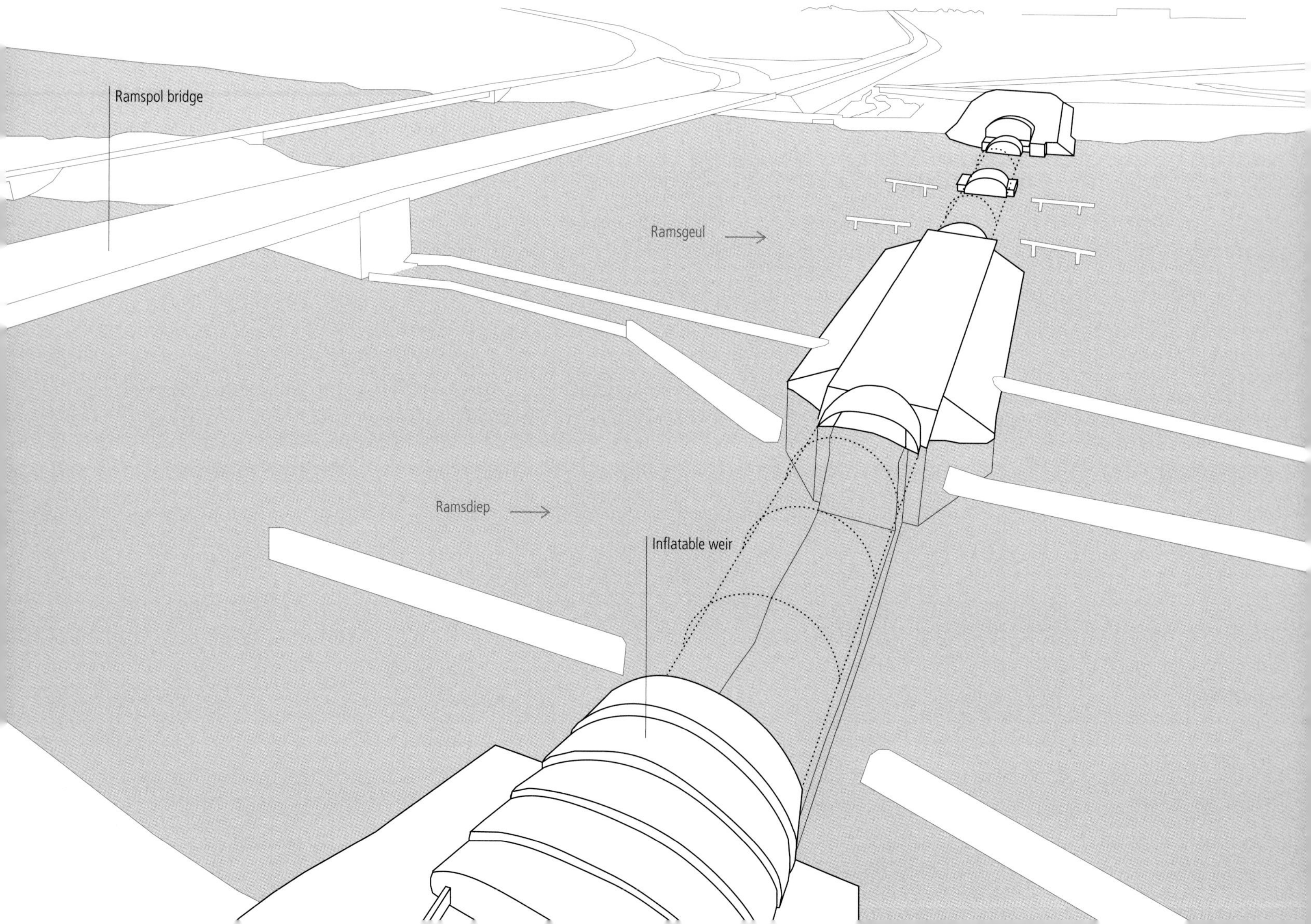

R

Pannerden Partition, Pannerden

An unsuspecting hiker in the floodplains near the village of Pannerden might easily misconstrue this 175-metre-long concrete construction for some kind of artwork by a landscape artist. But the dozens of concrete rectangles that span across the winter dike actually form the Pannerden Partition. The outermost rectangles are partially or completely filled with concrete slabs, while the middle ones are open. On the winter dike side, the partition is bounded by a section of the dike with basalt blocks, and on the canal side by two concrete triangles.

Near the partition, the Rhine splits into the Waal and the Pannerden Canal, which was dug as a defensive structure in 1707. Two thirds of the Rhine water goes to the Waal, the rest to the Pannerden Canal which connects the Bijland Canal (a part of the Upper Rhine that was dug in 1776) to the Lower Rhine.

The eighteenth-century canals were not the first interventions at the Rhine and Waal bifurcation. As early as the first century A.D. that location was already home to the 'Drusus Dam,' named after the Roman military commander Nero Claudius Drusus. It caused more water to flow through the Rhine and less through the Waal, thus making the Rhine more navigable, and creating a natural border of the Roman Empire.

The partition is flood defence technology at its simplest. If the floodplains are dry – which they are most of the year – it lies idle. Only when flood plains are flooded during periods of high water does the partition do its work, almost automatically. Depending on the expected height of the water, the concrete grid is partially or completely filled with the concrete slabs. These ensure that the water from Bijland Canal is distributed over the Waal and the Pannerden Canal in the same proportion as it is during low-water levels.

 Pannerden Partition, Pannerden 2014
Design Borm & Huijgens and Bureau de Beaufort

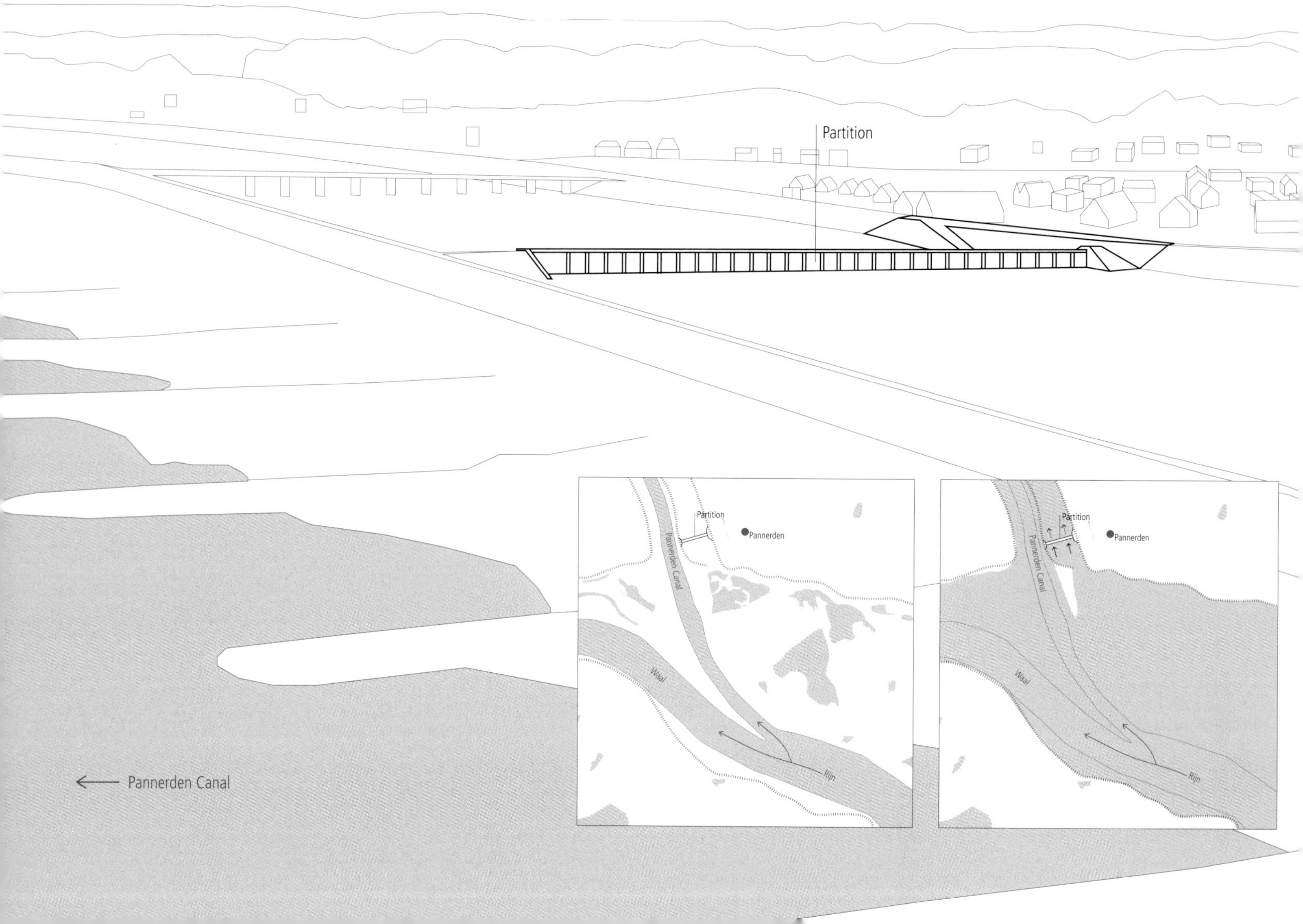

Weir in the Dommel river, Sint-Michielsgestel

152 Weir in the Dommel river, Sint-Michielsgestel 2016
Design Jan and Bram Taks

High Water Channel, Veessen

The Hoogwatergeul (high-water channel) between Veessen and Wapenveld, part of the Room for the River programme, covers an area that is eight kilometres long and between 0.5 and 1.5 kilometres wide. The channel was constructed to relieve the IJssel in the event of extremely high water levels, and includes an extensive system of almost parallel dikes, two bridges that are transverse to the new dikes, inlets and outlets at the beginning and end of the channel, and two pumping stations. There is also a new nature reserve and canoe route to the west of the westernmost new dike. An area for peatland birds has also been created within the new dikes.

No matter how awesome and extensive the changes to the landscape between Veessen and Wapenveld have been, in essence the Hoogwatergeul, just like the Pannerden Partition, is low tech. At extremely high water levels, the valve construction at Veessen is opened up, and the water flows under the new bridge into the undeveloped agricultural area, creating a channel that then connects to the IJssel at Wapenveld. This means that approximately 45% of the water is discharged in the IJssel, and the water level drops by about 70 centimetres. When the high water level in the IJssel has subsided, the two pumping stations empty the channel through the outlet at Wapenveld. It is expected that this will be necessary only once every eighty years.

154 High Water Channel, Veessen 2017
Design ZUS

TOLBRUG

Mills and Pumping Stations, Kinderdijk

The best-known mill corridors – which consist of two or more windmills in a row that lift the low-level water in a polder from a uniform low level to a uniform high level – are those at Kinderdijk. This village in the province of South Holland has eight windmills on one side of the Hoge Boezem, which move water from the Nederwaard polder to the basin, and on the other side eight windmills that do the same with the water from the Overwaard. The windmills of the Nederwaard are stone 'grondzeilers' (windmills that can be controlled from the ground) built in 1738, while those at the Overwaard are wooden grondzeilers dating from 1740. Near the 16 mills is another mill meant for draining the Blokweer polder, and two more mills to drain the Lekkerland polder.

Nowhere else in the Netherlands are there so many polder mills in one place. This is because in the 17th and 18th centuries, it was customary to demolish the mills that had dried out the lakes, such as De Schermer and De Beemster, if they were no longer needed for draining the polders. In addition, since the beginning of the 19th century, many mills have become superfluous due to the construction of pumping stations powered by steam, and later by electricity.

Kinderdijk also has pumping stations today. But the older steam-powered pumping stations built in 1868 were designed to increase the mills' water transfer capacity, and the pumping stations that replaced these mills only appeared in 1950. By that point, the mills at Kinderdijk had already grown into the definitive symbol of the Netherlands' 'battle against the water,' which is why they have remained. In 1997, they were placed on the UNESCO World Heritage List.

160 Mills and Pumping Stations, Kinderdijk 1795, 1972, 1995
Design —, P.A.C. de Bruin, Ir. G.N. Kok

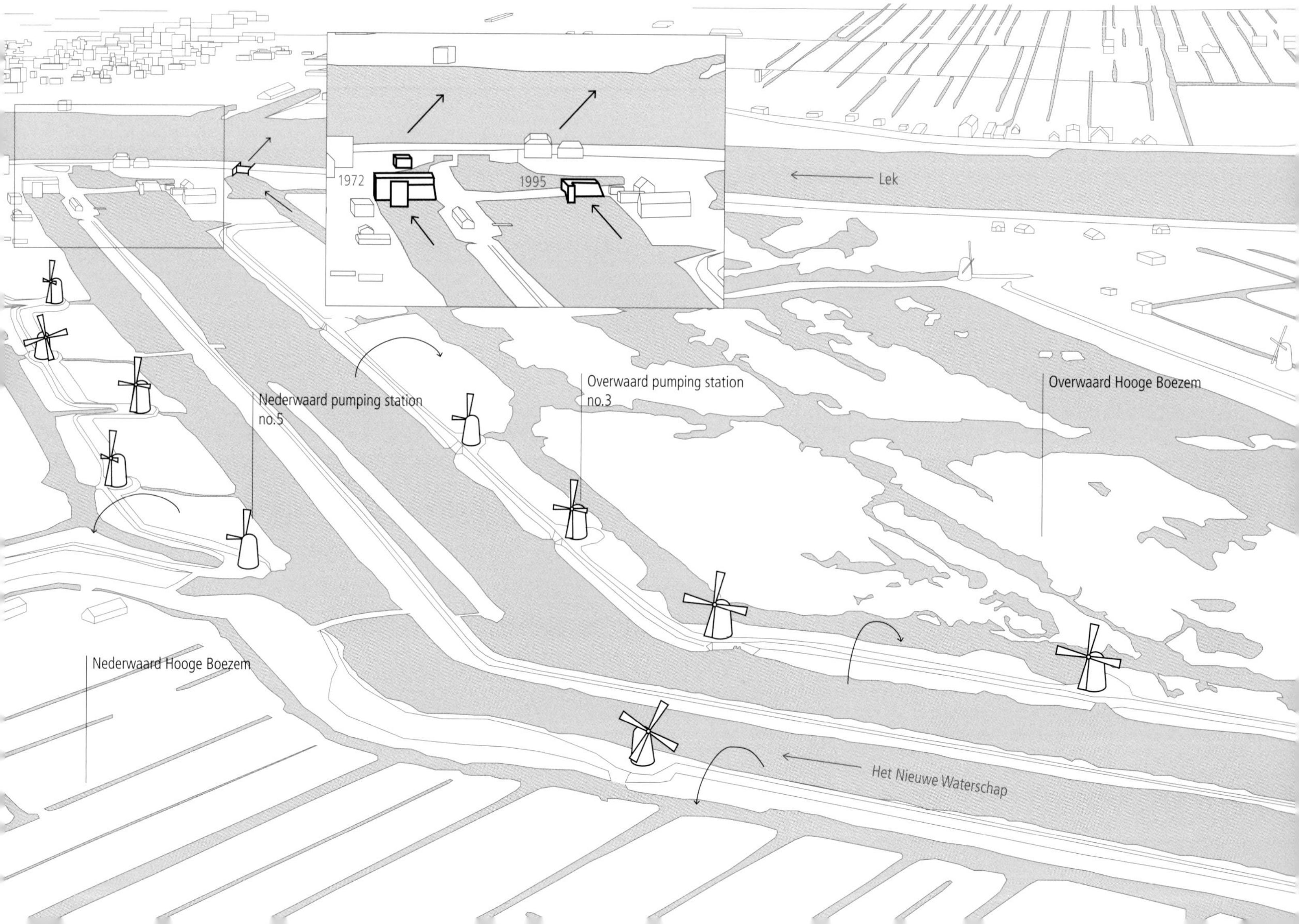

De Cruquius Pumping Station, Cruquius

In no other building is the Dutch battle against the water exemplified as well as it is in the De Cruquius pumping station, one of three such stations that pumped the Haarlemmermeer dry. This pumping station is named after the hydraulic engineer Nicolaus Kruik (Latinised: 'Cruquius,' 1678-1754), who made plans in the eighteenth century for the reclamation of the Haarlemmermeer, which had been steadily increasing in size due to the 'waterwolf' phenomenon (the tendency of lakes in low-lying peat lands to enlarge due to flooding and erosion). It is a neo-Gothic bastion that lies in the middle of the water. What makes it unusual is that this bastion is not a defensive work, but instead it battled against waterwolf using eight iron balancing arms (each weighing 10,000 kilograms) that stick out of the walls. Pumps are attached to the arms. Powered by the world's largest remaining steam engine, each stroke of the arms lifted 64,000 litres of water five meters up, where it was poured out onto the wooden stortvloer (landfill floor) that surrounded the engine room. From there, the water flowed through the locks on both sides of the boiler house into the ring canal.

Much as he did with the other two Haarlemmermeer pumping stations, the Lijnden and the Leeghwater, engineer Jan Anne Beijerinck (1800-1874) also designed the De Cruquius in the neo-Gothic style. Why Beijerink chose for neo-Gothicism is not clear, but it may have been because this style was considered modern and rational at that time, and therefore suited to the industrial era.

Unlike the Haarlemmermeer's other two pumping stations, De Lijnden and De Leeghwater, the De Cruquius pumping station has never been modernised. From 1912 onwards, the De Cruquius became a reserve pumping station, and in 1933 it was taken out of service. A year later, it became a museum. In 2002, the De Cruquius was again set in motion, after a restoration that lasted twenty years.

166 De Cruquius Pumping Station, Cruquius 1849
Design J.A. Beijerink

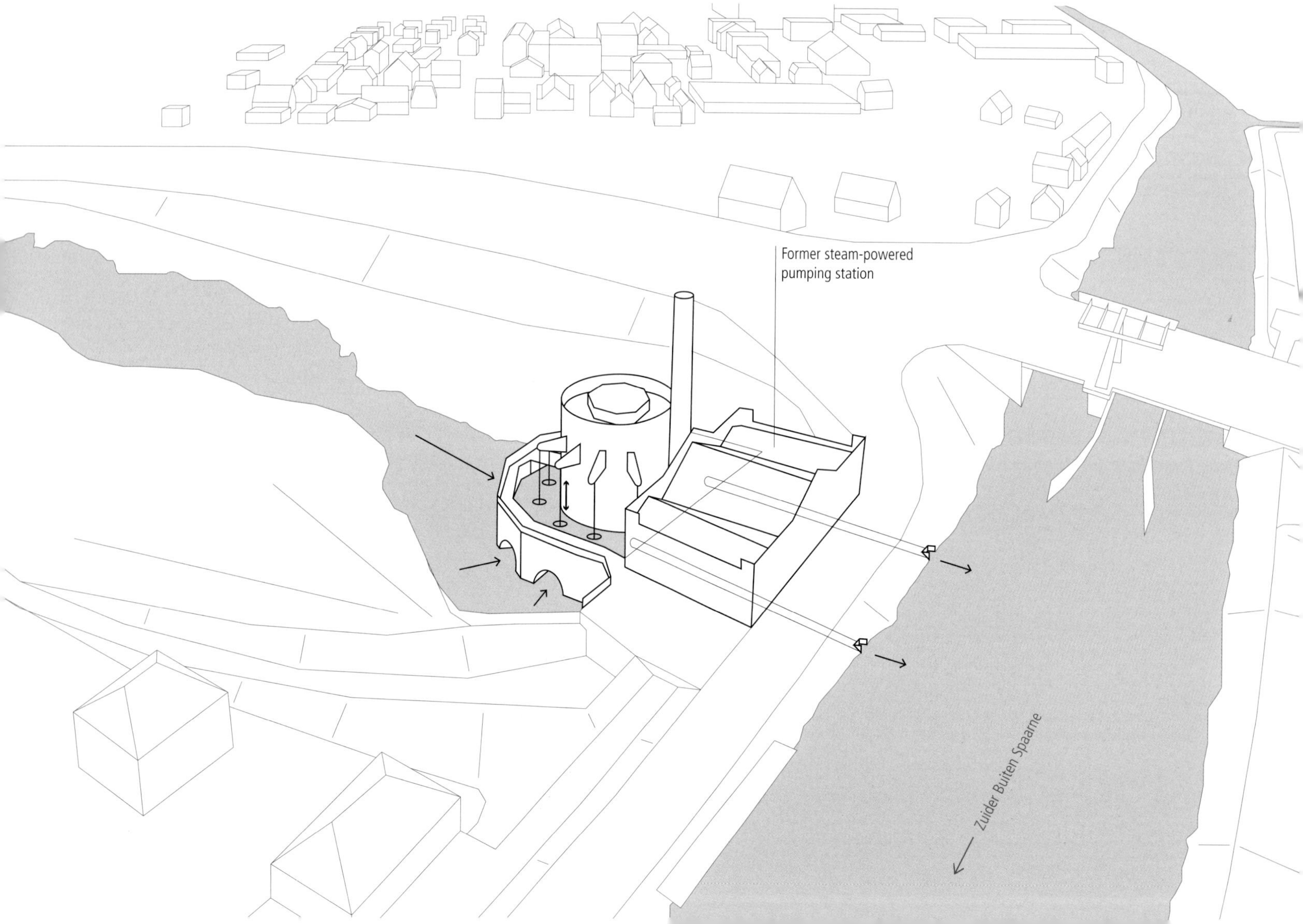

Cruquius
50
Cruquius
50
P

Basin Pumping Station, Rotterdam

170 Basin Pumping Station, Rotterdam 1859, 1891
Design O. Fynje and G.J. de Jongh

172 Wouda Pumping Station, Lemmer 1920
Design ir. D.F. Wouda

Wouda Pumping Station, Lemmer

The Ir. Wouda Pumping Station in the Frisian town of Lemmer functions as a dam in the Stroomkanaal, near the IJsselmeer. The largest steam-powered pumping station in the world, it was built between 1915-1920 to discharge the basin water of the Frieze Boezem (a system of canals, lakes and waterways that are connected to each other) into the Zuiderzee (which is now the IJsselmeer).

This well-preserved pumping station, consisting of a concrete substructure in the Stroomkanaal with atop it a simple machine hall for the motors that drive the eight pumps, was used until 1966. It is now only used a few times per year on average in cases of exceptionally high basin levels in Friesland.

For his design of the pumping station, the eponymous architect Dirk Frederik Wouda (1880-1961), who was the chief engineer of the Provinciale Waterstaat van Friesland (the Provincial Water Agency of Friesland), took a close look at the work of H. P. Berlage, the most influential architect of the Netherlands in the early twentieth century. With their natural stone lintels, as well as the brick tympanen with natural stone accents, the facades are thoroughly Berlagian. The spirit of Berlage can also be found in the interior, for example in its strips of blue glazed tiles.

In 2011, the Wouda pumping station, which was placed on the UNESCO World Heritage List in 1998, and is one of the Top 100 Rijksmonumenten (Dutch national monuments), was expanded to include a visitor centre designed by Jelle de Jong architects. It consists of a long steel beam on legs in the middle of a pond. A high footbridge and a scaffold between these legs provide access to the centre.

Binnenvree Pumping Station, Coevorden

174 Binnenvree Pumping Station, Coevorden 1928
Design Nederlandse Heidemaatschappij

Mill, Gerkesklooster 1923
Design —

The 18 metre-high windmill, or 'American Windmotor,' model 'Hercules Metallicus,' was imported and placed on the Lauwers River.

Lely Pumping Station, Medemblik

Rijkswaterstaat's waterworks are not by themselves anything more than engineering structures in the architectural sense. In the first decades of the twentieth century, this issue was the subject of lengthy discussion. With the pumping stations and locks becoming larger and larger, protectors of the landscape pleaded for them to be better embedded in the old Dutch landscape. Architects' circles sounded the call for waterworks that were not only functional, but also 'beautiful.'

Some water management engineers reacted resentfully to this call for adaptation and beauty. The head of Rijkswaterstaat's bridge department W.J.H. Harmsen, for example, proved to be a believer in the functionalist dogma that something that works well is naturally beautiful. In his view, the able and cost-conscious constructors had no choice but to design bridges that fit nicely into the Dutch landscape.

But in the 1920s, S. de Clercq, chairman of the Bond van Nederlandsche Architecten (Association of Dutch Architects) and member of the Zuiderzeeraad (Zuiderzee Council), repeatedly pleaded for architects to also be involved in the design of the pumping stations and other waterworks. And he did so successfully: at the end of the 1920s, Rijkswaterstaat awarded Dirk Roosenburg the contract for the Lely and Leemans pumping stations that were built for the reclamation of the Wieringermeer, which was the first of the IJsselmeer polders. From the Lely Pumping Station, named after ir. Cornelis Lely (who as the Minister of Waterstaat was the driving force behind the Zuiderzee works), Roosenburg made a brilliant composition of rectangular volumes in which he used windows as ornaments. Even after the reclamation of the Wieringermeer, the Lely Pumping Station (which bears resemblances to W.M. Dudok's town hall in Hilversum) remained in function, and still does today.

176 Lely Pumping Station, Medemblik 1929
Design ir. D. Roosenburg

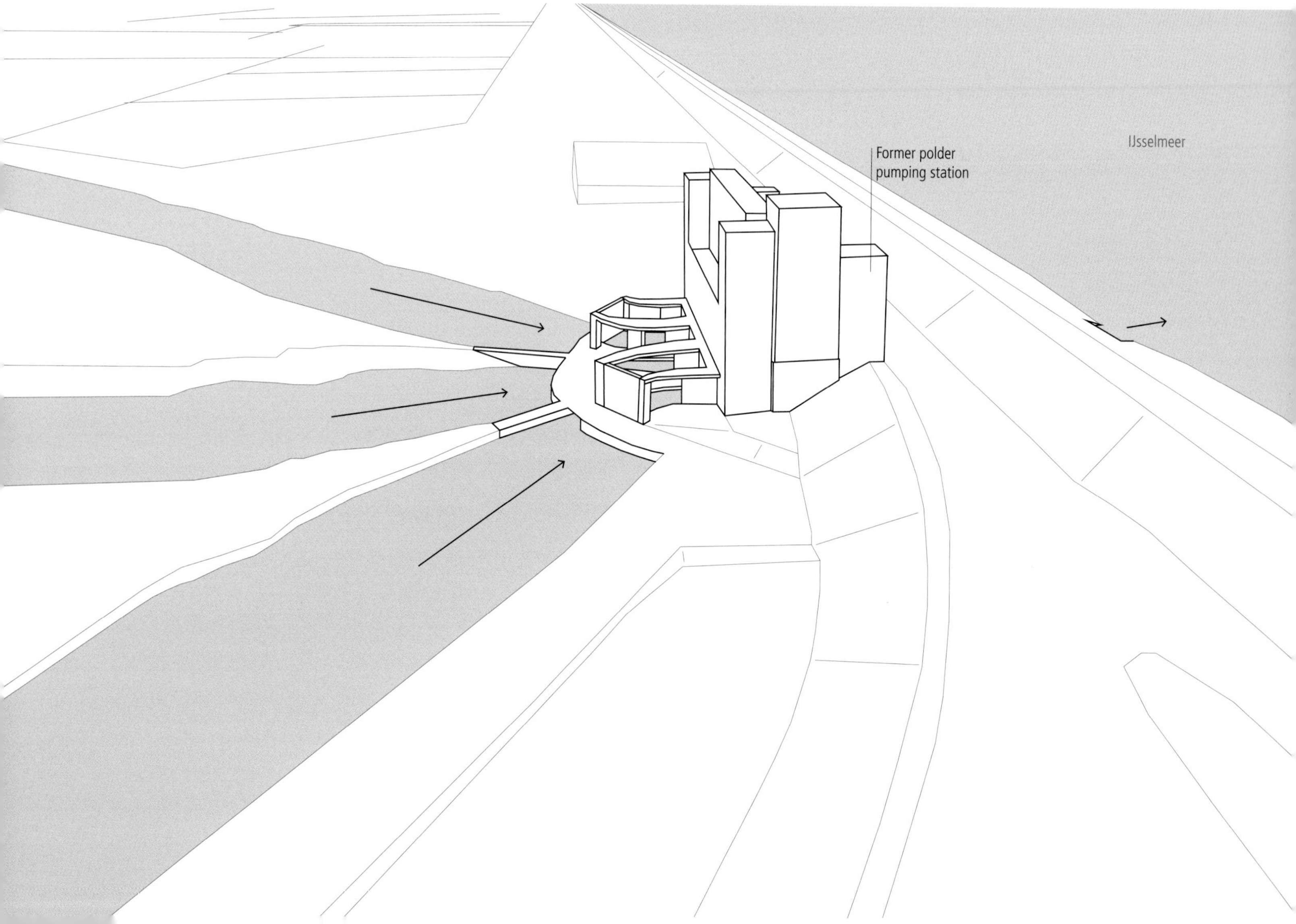

LELY

LELY

Veneriete Pumping Station, Genemuiden

The Veneriete Pumping Station, dating from 1961, replaced the steam-powered Mastenbroek Pumping Station (built in 1856), and now stands next to the Kamperzeedijk. This pumping station drains a large part of the Mastenbroek polder by using three screw-pumps to pump up the polder water from a lake to the Veneriete, which is a drainage canal that ends in the Zwarte Meer.

The most remarkable thing about the Veneriete Pumping Station is the transparency of its upper section. Passers-by who cycle or drive over the two-section bridge along the provincial N760 road can see, through the large window with round corners, the electric motors that drive the screw-pumps.

The pumping station, designed by Willers Architectuur, is a fine example of the Shake Hands style. In this style from the reconstruction period, brick traditionalism is united with the concrete and glass of Nieuwe Bouwen. Many original details are still present in the Venerite Pumping Station, such as the slightly flared stairs on the dike, with their typically reconstruction period railings, and the cheerful, slightly flapping roof edges.

180 Veneriete Pumping Station, Genemuiden 1961
Design Willers Architectuur

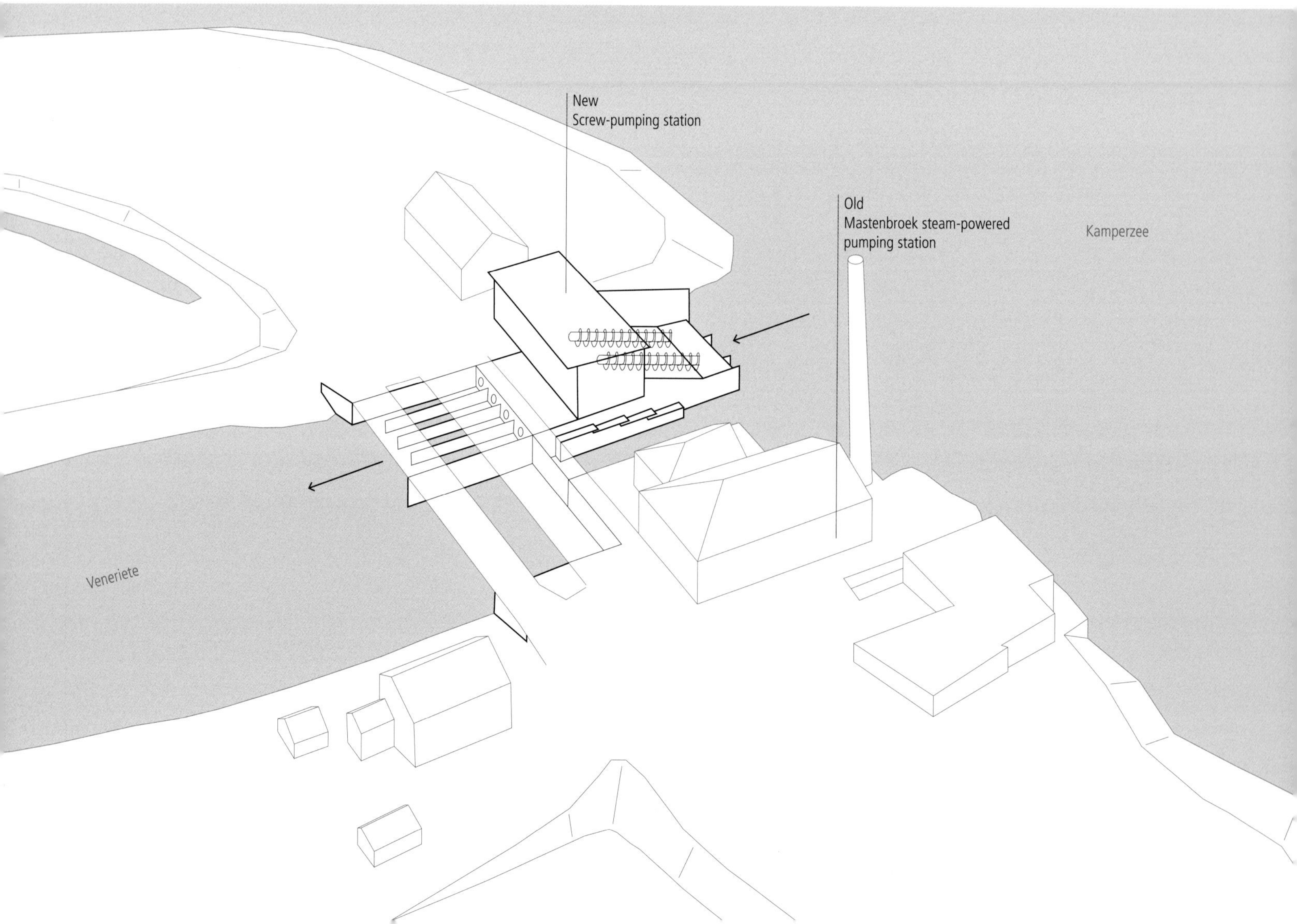

GEMAAL VENERIETE

De Laar Pumping Station, Gellicum

184 De Laar Pumping Station, Gellicum 1962
Design —

Abraham Kroes Pumping Station, Zuidplaspolder

The Abraham Kroes Pumping Station is a unique waterwork in the Netherlands. It is named after a dijkgraaf from the Zuidplas polder, which was reclaimed in the first half of the nineteenth century, and is the only pumping station to carry out two different types of draining. It not only brings water from the Zuidplas polder, where the lowest point of the Netherlands is located (6.76 metres below NAP) to the basin, but also discharges the basin water into the Hollandse IJssel, from where it flows further to the sea.

Like so many other lakes and ponds in the Netherlands, the Zuidplas was formed over the centuries by peat extraction. And again, like so many places in the Netherlands, it grew larger due to the 'waterwolf' phenomenon (see p.166). In 1825, the decision was made to impolder the Zuidplas. Fifteen years later, the lake was drained. Due to the settling of the peat soil, the ground level of the polder has since sunken further. In 1995 it had sunken so much that the Survey Department of Rijkswaterstaat found that the lowest point of the Netherlands was no longer where it used to be (in the polder Het Lage Land), but rather in a meadow along the Derde Tochtweg in the Zuidplaspolder, at 6.74 meters below NAP. Ten years later, the Zuidplaspolder was found to have sunken two centimeters more.

The Abraham Kroes pumping station was built during the years 1968-1971 on the thirteenth-century dike along the Hollandsche IJssel river, near the Snelle Sluis of 1829, the first waterwork built for the impoldering of the Zuidplas. Two architects were commissioned for the design of this double pumping station, J. van den Boogaard and J. de Haan. The trouvaille of the Abraham Kroes Pumping Station is that the drainage pipes of the basin water have been integrated in a theatrical, 80-metre-long ramp that runs from the Schielandse Hoge Zeedijk West over the ring canal to the entrance of the pumping station.

For the pumping station itself, the architects used extremely simple means. From the front facade of the two staggered boxes built against the dyke of the ring canal, they made a De Stijl-like composition using a horizontal strip of six windows, a door and a chimney, and brick walls. On the polder side, the facade is even more sober, with three vertical strips of square windows and strips in light and dark.

186 Abraham Kroes Pumping Station, Zuidplaspolder 1971
Design J. van den Boogaard and J. de Haan

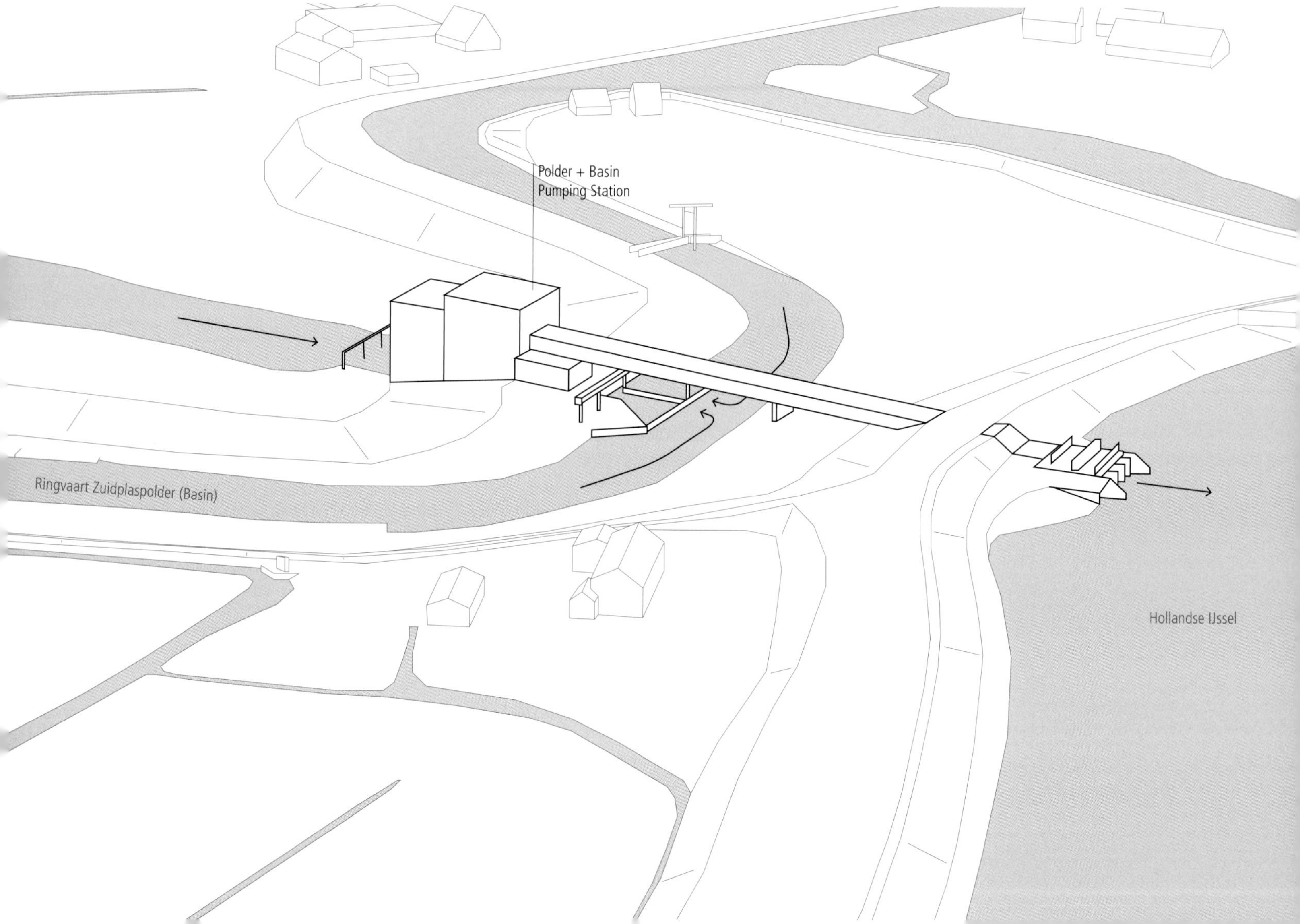

5

Schieland en de Krimpenerwaard

De Helsdeur Pumping Station, Den Helder

190 De Helsdeur Pumping Station, Den Helder 1972
Design ir. Postel, Tauw

De Helsdeur is one of the largest pumping stations in the Netherlands, with a draining capacity of 3,000 cubic metres per minute.

GEMAAL DE HELSDEUR

BONUS

 De Leyens Srew-Pumping Station, Zoetermeer 1975
Design —

De Leyens Screw-Pumping Station, Zoetermeer

When the De Leyens vakgemaal or vijzelgemaal (screw-pumping station) was completed in 1977, little could be seen of the residential area that now exists there. The construction of Zoetermeer, a new city near The Hague in the Zoetermeerse polder (which was reclaimed in 1771), was still in full swing at the time. Later, when the De Leyens district was completed, the pumping station was one of the few places in the Netherlands where passers-by could see an operational 'screw of Archimedes' (a screw-pump in a shaft). Three large screw-pumps were located next to each other on the lower side of the building. They lifted the water from the Zoetermeerse Singel (5.8 metres below NAP) 1.8 metres up to the Broekwegwetering.

In 1992, the middle screw-pump was replaced by a pump that could increase the pumping capacity. Because the De Leyens Screw-Pumping Station and the two remaining screw-pumps made a lot of noise, they were covered in 2016, meaning that nothing of this attraction can be seen today.

An interesting detail about the pumping station's simple building, whose designer is unknown, is that both the two dense walls and the flat roof extend a bit further than necessary. From a distance, this makes it resemble the architecture of De Stijl from a century ago.

In 2017, the De Leyens Screw-Pumping Station underwent a thorough refurbishment, based on a design by Kruunenberg architects, whereby the screw-pumps and the pump were replaced. The two screw-pumps are now surrounded by a round casing.

Schilthuis Pumping Station, Rotterdam

Over the centuries, Rotterdam has needed to change its drainage system several times. This happened for the first time in the first half of the eighteenth century, when the peat lakes in Schieland were drained, and the basin river the Rotte was not large enough to discharge the increasing amount of water, which is why a second basin was constructed in 1742. New basin channels were dug later, the Lage Boezem and the Hoge Boezem.

In the early 1970s, Rotterdam's drainage system had again become insufficient. There were no new basins dug this time, but instead the Schilthuis Pumping Station was built on the Oostplein (at the end of the Hoge Boezem), which had since become part of the city centre. For the pumping station, the designers from DHV, Pouderoyen, and Sramota created a modest and dynamic enclosure with sloping side facades. The facade on the Oostplein is particularly striking: with its long horizontal slats, the top side forms contrasts with the vertically articulated facade of the substructure.

Many people who see the pumping station, named after the twentieth-century dijkgraaf U.G. Schilthuis, will wonder where the water pumped out of the Hoge Boezem goes; via pressurized pipelines, it is transported underneath the Oostplein to the Boerengat, about 200 metres away, which itself is connected to the Nieuwe Maas.

196 Schilthuis Pumping Station, Rotterdam 1977
Design DHV, ir. Pouderoyen and ir. Sramota

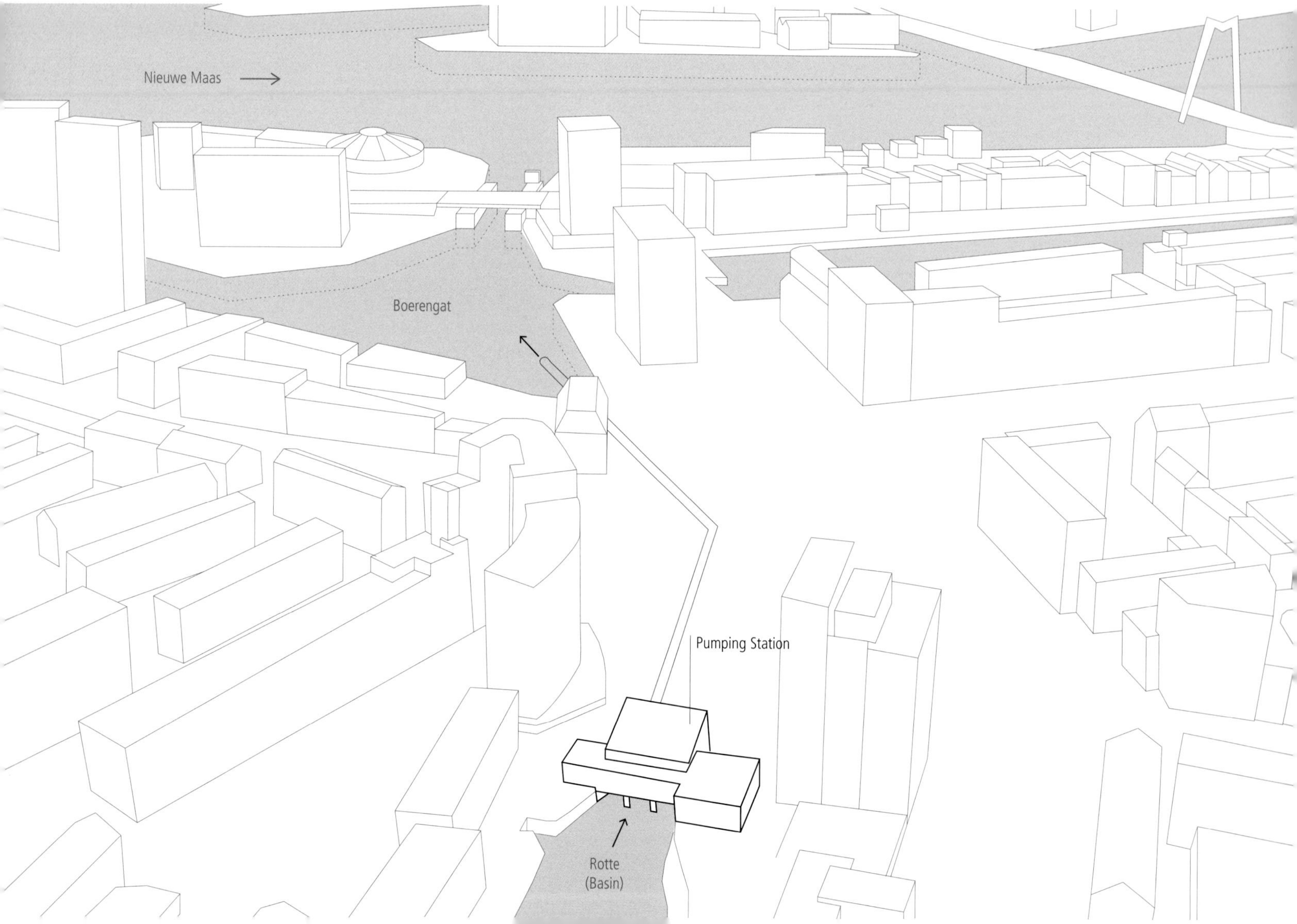

GEMAAL
410-42

Hoogheemraadschap van
Schieland en de Krimpenerwaard

GEMAAL MR. U.G. SCHILTHUIS

Kostverlorenzijl Pumping Station, Zwartsluis

200 Kostverlorenzijl Pumping Station, Zwartsluis 1992
Design Haskoning

OSTVERLORENZIJL

De Ronde Hoep Pumping Station, Ouderkerk aan de Amstel

202 De Ronde Hoep Pumping Station, Ouderkerk aan de Amstel 1913, 1995
Design W.C de Wit and K. de Wit

Pumping Station, Waterland

204 Pumping Station, Waterland 1999
Design Jord den Hollander

In Waterland, the area north of Amsterdam, a number of these identical small pumping station were placed.

Joh. Veurink Pumping Station, Krimpen aan den IJssel

206 Joh. Veurink Pumping Station, Krimpen aan den IJssel 2003
Design →

Pumping Station, Nieuwersluis

208 Pumping Station, Nieuwersluis
Design —

Pumping Station, Leuvenheim

Pumping Station, Leuvenheim 1964
Design —

Onnerpolder Pumping Station, Onnerpolder

The Onnerpolder Pumping Station is a 'stairway to heaven.' In this case a bird's heaven, because the staircase leads to the roof of the pumping station (entirely made of concrete), where many birdwatchers now come to visit the nature reserve that the Onnerpolder has since become. Architectural experts will also be reminded of Casa Malaparte, the famous house of the Italian writer Curzio Malaparte (1898-1957), whose roof also consists of a wide staircase with a terrace.

The location of the pumping station is similar to that of the Malaparte house. Just as Casa Malaparte is located like a bunker on a peninsula (on the island of Capri in the Mediterranean Sea), so does the Onnerpolder pumping station resemble 'a bastion,' as Onix Architects themselves describe it, on the border of the Onnerpolder and Westerbroekmadepolder. Dikes and ditches come together here, and the designers say that the lines of the landscape have been extended to the pumping station, resulting in a knotty sculpture. All the elements of the pumping station, such as the culverts, deck cleaners, weirs and waste bins, have been integrated so as not to detract as disturbing appendices from the overall sculptural form. The angular pumping station has been extended with a concrete terrace along the water, with a water inlet on one side and a narrow beam on the other. The narrow beam extends over the water, and it thereby also serves as a bridge.

210 Onnerpolder Pumping Station, Onnerpolder 2004
Design Onix Architects

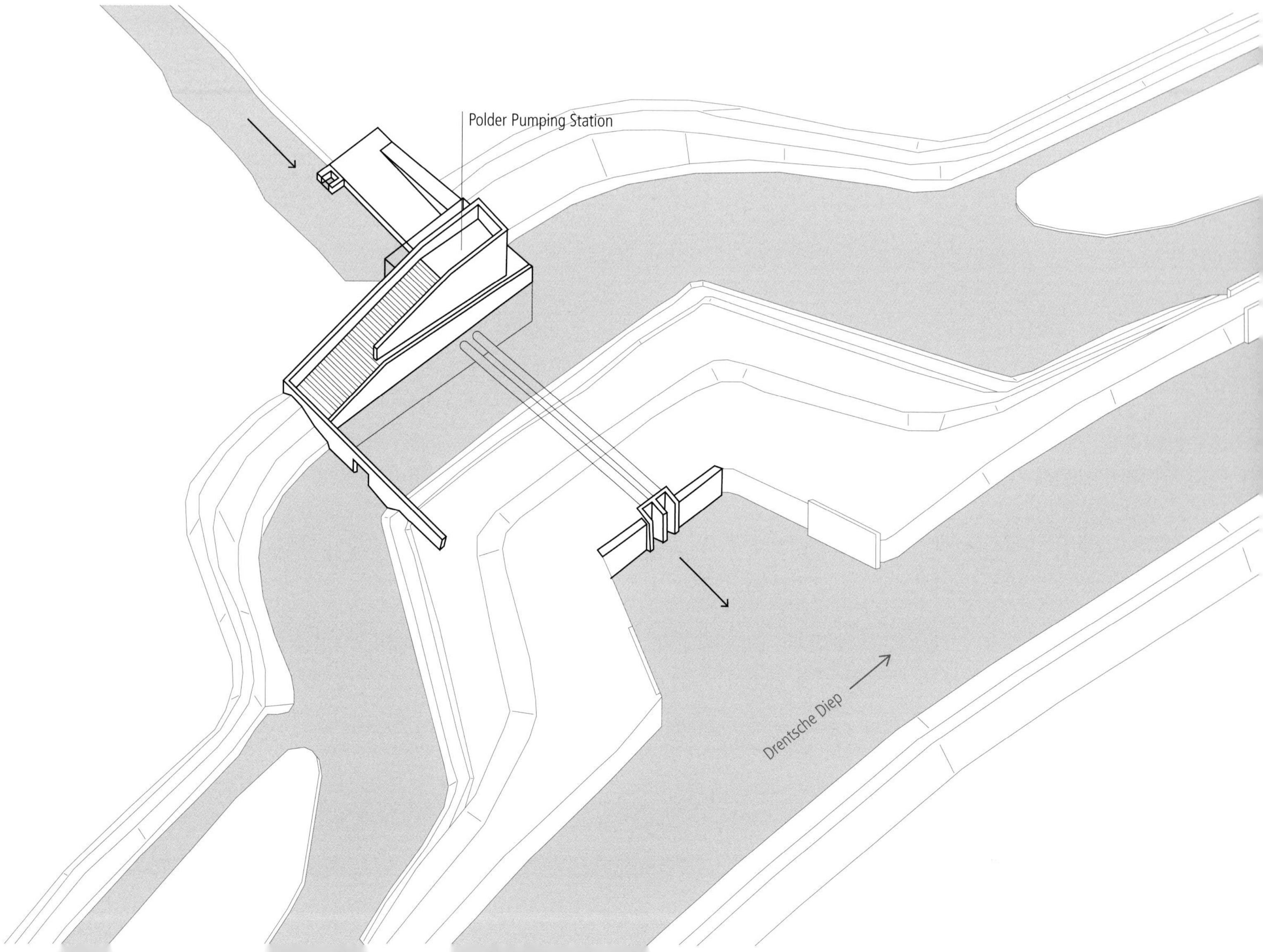

Lansingerland Pumping Station, Bergschenhoek

214 Lansingerland Pumping Station, Bergschenhoek 2007
Design Tauw GmbH, Oomen Architects

Kort Haarlem Pumping Station, Gouda

Kort Haarlem Pumping Station, Gouda 2011
Design Aletta van Aalst & Partners

Rijpwetering Pumping Station, Rijpwetering and Hoogeveen Pumping Station, Hazerswoude

In 2008, the Hoogheemraadschap van Rijnland (Rijnland District Water Control Board) decided that 'the recognisability of the structures in Rhineland should be increased.' To this end, a series of standard designs for pumping stations, with their own 'Rhineland' character, had to be developed. The designers of the pumping stations were commissioned to make the new pumping stations look more like machines than buildings. To enhance that sense of recognition, the Rheinland Water Board increased the construction budget of the new pumping stations by five percent, or 150,000 euros per year.

The Groenendijk polder pumping station in Hazerswoude, designed by Aletta van Aalst & Partners, is one of the stations built as part of this project to increase the recognition. This small 'vijzelgemaal' (screw-pumping station) is genuine 'architecture parlante': in the concrete sculpture, you can see a petrified wave that spans across the dike. Glass facades in the upper part of that wave show the motors that drive the screw-pump, which is hidden underneath a segmented hood that is easy to dismantle.

Aletta van Aalst & Partners also designed new sculptural superstructures for the three polder pumping stations in Rijpwetering, which consist of segments. The Gemaal Drooggemaakte Hertogs-, Akkersloot- en Blijverspolders pumping station is an example of unadulterated postmodern Las Vegas architecture. On the dike is a square sign with the name of the pumping station printed on it, surrounded by a picture frame. Behind that is a hood that starts at the top of the dike like a rectangular box, then takes on a flowing shape, and finally ends in flat segments at the bottom of the dike. This pumping station is also 'talking architecture' reminiscent of water: it is as if a larger bucket of water is being emptied above the dike.

In 2011, when sculptural pumping stations were built in addition to the one in the Groenendijk polder, including in Haarlem and Gouda, the Rijnland District Water Control Board noted that 'it is difficult to coordinate between architectural design and functional requirements.' This had led to delays in the construction of new pumping stations, and the additional construction costs were much higher than the estimated five percent. That is why Rhineland decided to cut back on the new generation of pumping stations.

216 Rijpwetering pumping Station, Rijpwetering and Hoogeveen Pumping Station, Hazerswoude 2014, 2011
Design Aletta van Aalst & Partners

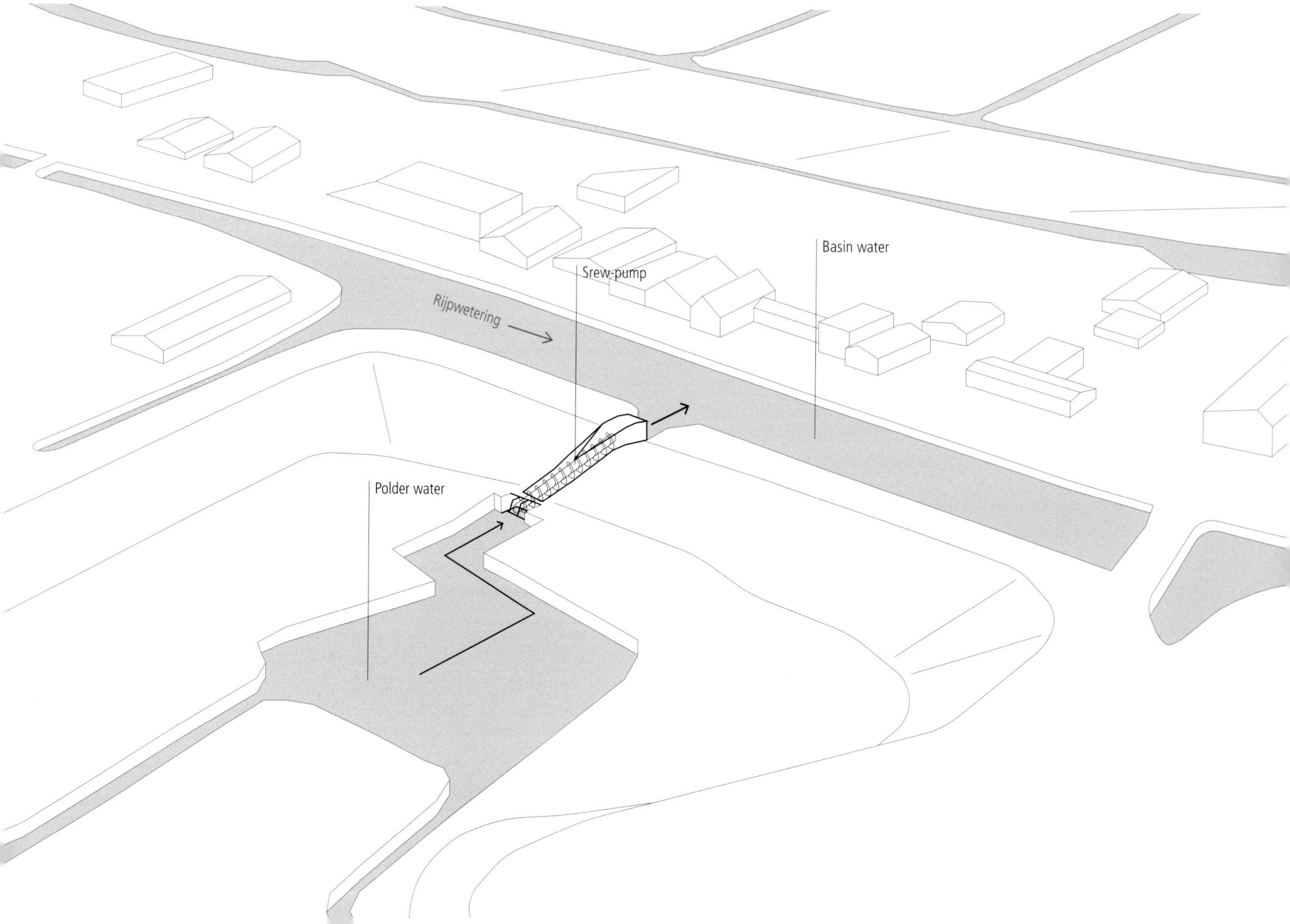

GEMAAL POLDER GROENENDIJK

GEMAAL
DROOGGEMAAKTE
AKKERSLOOT-,
HERTOGS- &
BLIJVERPOLDER
Hoogheemraadschap van Rijnland

Weijerswold Pumping Station, Coevorden

218 Weijerswold Pumping Station, Coevorden 1966
Design —

Eendragtspolder Pumping Station, Zevenhuizen

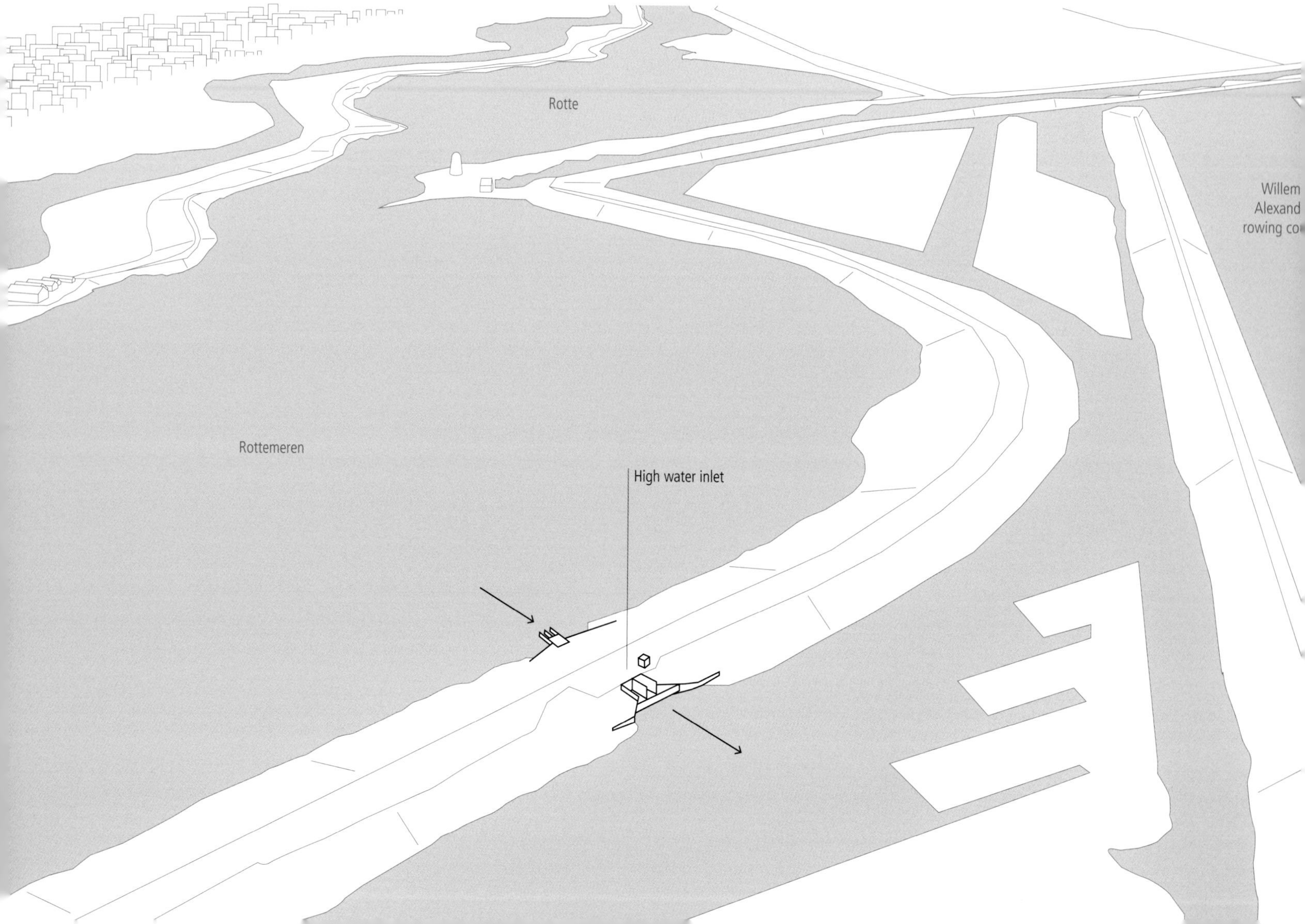

Noordwaardpolder Pumping Stations, Biesbosch

The twelve new pumping stations in the Noorwaardpolder, an area of almost 4,500 hectares north of the National Park De Biesbosch, were built as part of the Room for the River programme. The Noordwaard, on the Merwede river, is one of the largest of the 30 areas in the Netherlands that have been designated to give rivers more space, so as to reduce the risk of flooding in the river area. To this end, the Noordwaard dike along the Merwede was lowered to two metres above NAP, so that the former agricultural area is now flooded several times a year in cases when the water level in the Merwede is high.

To make the Noordwaard more than just an overflow, Rijkswaterstaat organised a competition in 2009 to make the de-poldered area suitable for recreation. With a design created by West 8 and IPV Delft for twelve pumping stations and 30 bridges that would also make it possible to walk and cycle through the area during high-water levels, the Combinatie Noordwaard won the competition.

The twelve pumping stations play a role in the cycle route. The simple rectangular buildings serve as beacons in the de-poldered Noordwaard, and from several pumping stations, the roofs serve as viewpoints, accessible via stairs and/or walkways. The various types of bricks that make up the facades of the pumping stations come from around 50 farms and other buildings that were demolished for the de-poldering. In this way, the new pumping stations connect architecturally with the pumping stations in the neighbouring Biesbosch, which have always been simple, traditionalistic buildings.

222 Noordwaardpolder Pumping Stations, Biesbosch 2015
Design IPV Delft and West 8

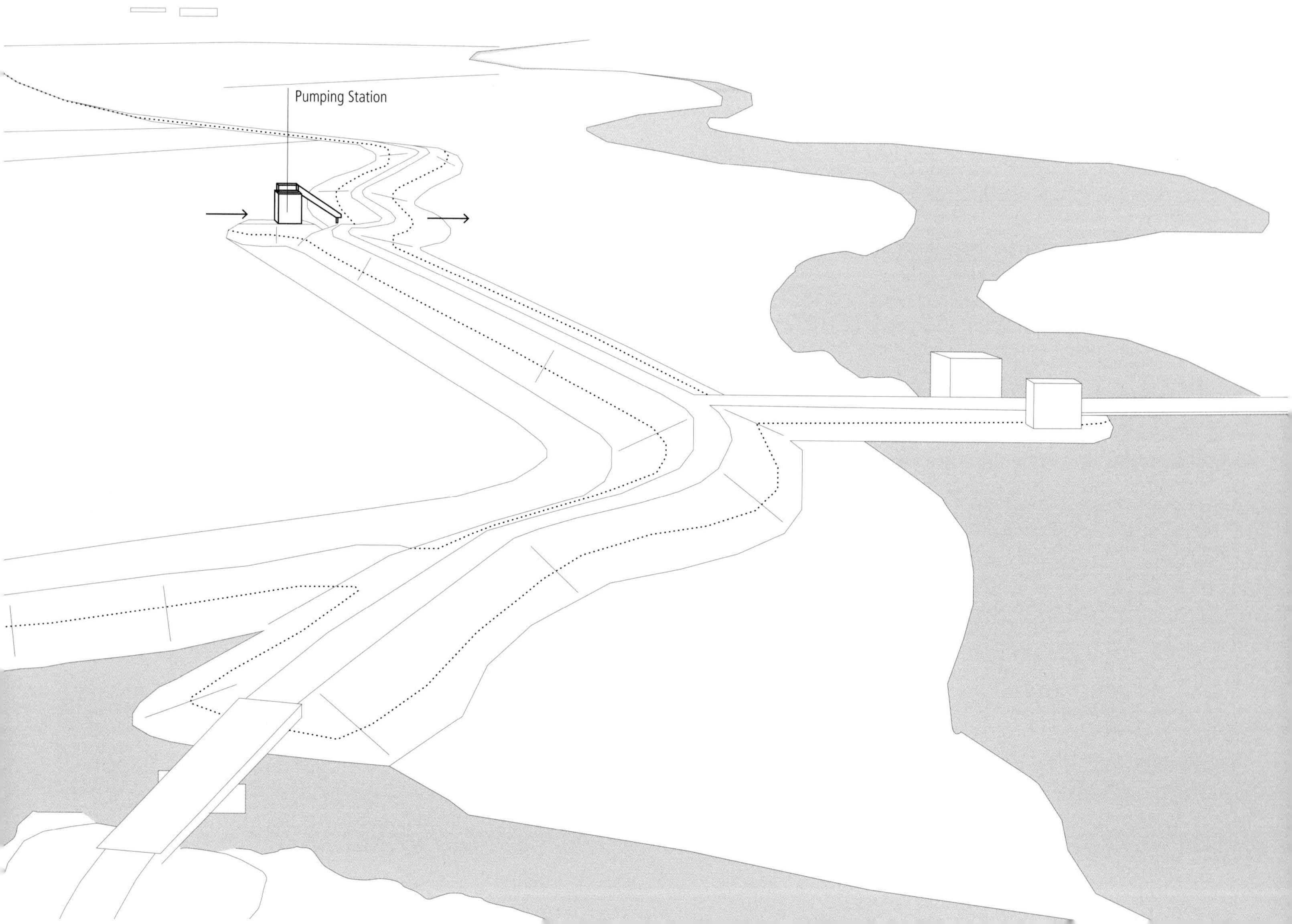

Essay

The design of built flood defences has a very long tradition, which illustrates that our topography owes its ongoing existence to active water management. The flood defences are the 'prostheses' with which our country remains dry, as it would otherwise be largely made of water. The interesting thing about the collection of waterworks is that a certain kind of Dutch constructivism, produced entirely by the public sector, has been created from this almost uniform type of utilisation. This collection was built over the course of the twentieth century, and that is the main reason for a comprehensive assessment of its effectiveness, in light of the fact that an important part of this collection will soon be reaching the end of its lifespan, in addition to the fact that we have to redesign the water management system in a climate-adaptive way.
It is important to think in advance about which architectural conditions and ambitions this challenge should be imbued with. How do the fragmentation of commissioning practices and the broader objectives of water management (ecology, sustainability, energy neutrality, recreation) affect the architectural specifications of the new waterworks? That is because the approach and working method of the Delta Programme has given a huge boost to long-term thinking about the future of water management. On the basis of KNMI's climate models, 2100 suddenly seems a lot closer and more imaginable than it was ten years ago. Recent insights into climate change development show that extremes that were initially estimated to be realistic in 2100 might actually become the reality 50 years earlier.

Without losing sight of the historical context, that is why the book **Waterwerken in Nederland** focuses on the twentieth and twenty-first centuries, with an occasional excursion to some historically formidable works. What factors were decisive in the last century for the construction of waterworks, their design, their location, and the mechanisms that they used (and still use) to carry out their work? What were the social, technical, and physical-spatial influences on the oeuvre as it was formed in the last century? How did we deal with the tension between relying on tried-and-tested techniques and tactics, while at the same time innovating and anticipating new social insights?
And what are the new requirements and conditions for waterworks in the twenty-first century? How are influences such as unstable climate forecasts, the pursuit of ecological and landscape mitigation, and redistributed responsibilities materialised?

These questions were raised during two expert meetings that were organised for this book. The first meeting, partly based on the photographed waterworks, looked for an interpretation of the 'organizing principles' behind the creation of the waterworks in the twentieth century. The second expert meeting examined how these organizing principles seem to be developing in the twenty-first century. The repercussions of these two meetings laid the foundations for this essay, for which several interviews were also held.
A number of the statements made in the expert meetings and interviews have been reproduced here (almost) literally. These powerful quotes speak for themselves, and through their editing, they reinforce the below text.

This essay meanders along a line of development in which the role of waterworks in the water system is explained in three phases. The first phase looks at the past; the waterwork as a 'victory,' a technical adaptation of the landscape system. Think for example of the Afsluitdijk and the large pumping stations. In the second phase, namely the present, the waterwork corrects the natural system, but at the same time mitigates the negative effects of doing so, for example the construction of the Oosterschelde storm surge barrier. In the third phase, the near future, the waterworks will begin to cooperate with the properties of the system. The waterworks are becoming more integrated, both in terms of their functional programme as well as the integration of their design with maintenance and management.

Technical Tradition

Thanks to the invention of techniques such as culvert valves, about two thousand years ago, it became possible to live in places beyond the higher-lying areas. The need to be able to manage water flows increasingly well, the development of the steam engine, and the use of new materials such as concrete and steel greatly increased the

number of waterworks. Nevertheless, disasters proved unavoidable, such as the massive floods of 1916, 1926, and 1953. In response, several unprecedentedly large and unique projects were undertaken in the twentieth century. The Afsluitdijk was built to contain the elusive Zuiderzee, and the Delta Works were built to make the south-west of the country safe once and for all. Likewise, the Room for the River programme was initiated in 2007 to help deal with the flooding threat posed by the rivers, and to help capture the excess water.
The construction of waterworks is based on a long tradition that trusted, and still trusts, in proven techniques. Many of these were developed from a basis of practically: there was an overview of what was needed, and where a waterwork would need to be placed. In principle, the ability to read the landscape was very much available. Where do you build the structure? Mills, followed later by pumping stations, were placed at more or less the same spots, even though this was not always logical for the larger area, which increasingly powerful pumps were capable of draining. The development of wind, steam, diesel, and electric drainage did not ultimately lead to another design of the polder landscape. The guiding idea was to consistently think in terms of a single object, and a single place. And that is still largely the case. Scaling up could have led to the waterworks being positioned differently. In civil engineering, innovations have emerged from improved modelling, more sophisticated calculation methods, new materials, and new construction methods. The invention of the steam engine, which made it possible to use less manpower to generate more and more pumping power, is one such innovation. The construction of the Afsluitdijk was enabled by larger excavator grabs. A second factor is the development of the computer, which was actually crucial for the Delta Works. Without this technology, these works would not have been possible. The impetus was already given by Hendrik Antoon Lorentz (1853-1928), who formulated a wave equation for the Afsluitdijk using a Millionaire, the first calculator that was capable of multiplying, and which would also later be used in the calculations for the Delta Plan.

Tjalle de Haan (Association of Waterbuilders / Vereniging van Waterbouwers and former Rijkswaterstaat employee) notes three main developments in this technical tradition. 'The first is the development in the constructive approach; elements made from zinc evolved from brushwood, and reed evolved to plastic fabric. This is all underwater, so you don't see it, but you do see it on the dike banks, where basalt was replaced by concrete elements or by asphalt, or by penetrated rock fill. The second development is the computer. In the past, we needed to look for analytical solutions to mathematical problems, but now those are becoming numerical solutions. You can handle almost infinite amounts of data with a laptop, whereas it used to take rooms full of human 'computers' (a former job title) who would sit there with their slide rules. The computer also led the way to the hydraulic crane, which is the third development: these cranes can now deliver work with centimetre accuracy, anywhere, even under water, or for very large stones, whereas in the past it was more a question of guesswork and just hoping that everything would be correct. The hydraulic cranes replaced the imprecise dragline, and the lifting capacities of floating cranes became greater and greater. We are now lifting elements that weigh of hundreds of tons. This was essential to our being able to construct the Oosterschelde storm surge barrier as a kind of prefabricated building.'

These technical innovations continue. 'Now we hydraulically press the piles instead of driving them,' said Hein Versteegen of Dura Vermeer during one of the expert meetings. 'The innovations in technology make more and more possible. Whereas you used to need to build a dike, now you can just build a sheet pile wall.'
'But we cannot expect too much from the executing contractors in this respect,' he added. 'Innovation is expensive and complex, and the whole world is watching. A builder is a service provider, in a highly competitive market. What we develop benefits others. So do expect us to keep developing smart logistics, and to learn from mistakes made in the past.' He notes that underground, considerably less has changed than above ground. 'The technology has remained the same,' he said.

Nardo Hoogendijk of Hollandia Infra adds to Versteegen's words. 'Innovation can also mean, for example, solving certain weight problems that arise in steel construction by using plastic. Yet this innovation also has a negative side to it, because steel is completely recyclable, and plastic is not, which means it is much less sustainable.'
He notes another technical innovation, which is partly the result of the arrival of the computer. 'The Maeslantkering was designed in a probabilistic way. Calculating in terms of 'failure factors' played a decisive role in the design process. For the first time in the history of waterworks, mankind was eliminated from the control process; the computer determines the control. Yet this is a development that is already receding. The infinite belief in the computer turned out to have its limits, after all. Nonetheless, you do still see a development in terms of remote control. People have become considerably less visible in the process.'

Landscape Influence
The design of waterworks is the product of key numbers and figures: general, standardised, and uniform criteria on the basis of which a solution is generated more or less automatically, and which then typologically adjusts itself to the existing repertoire. Nevertheless, local conditions (spatial, landscape) always play a role, which creates exceptions and idiosyncrasies. This leads to the question of the significance of specific circumstances, which have led to customisation in the hydraulic engineering activities of the Netherlands. Is the national collection of waterworks to be qualified as uniform, differentiated, or varied? Another question is the relative weight that local circumstances will or should be given in the future: where will the balance be reached between standardisation and customisation? How does that balance affect the design of the waterworks in terms of standardisation versus local factors?

Everything starts, as it always has, with the material that is available. The material, for example brick, has played a major role in the development of waterworks. Although later designers dealt more freely with the commissions, and even though they have started to build in a more artistic way – sometimes leading to a cacophony of materialization – in general it can still be said that there is a high degree of path dependence with waterworks. Choices that were made in the past in terms of the location and type of waterwork largely determine the new choices, and often exclude other options. Berno Strootman (Rijksadviseur voor de Fysieke Leefomgeving / State Advisor for the Physical Living Environment), mentions five factors that have played an important role in this respect, both technically and, above all, in the landscape, and that will also play an important role in future projects. 'The **first factor** was the rise of a number of brilliant visionaries, such as King William I, Jan Blanken, Cornelis Lely, and Johan van Veen, the founder of the Delta Plan. People who gave an enormous impetus to the development of waterworks over a period of decades. Think of the Zuiderzee works, for example. Disasters, as a **second factor**, have also been crucial. Think of not only the 1953 disaster, but also Room for the River and what preceded it, including the evacuation of a quarter of a million people in 1995. The **third factor** is the transition from control over nature to a game with nature. From a hard attitude to a soft approach. Towards one that moves along with nature. The renovation of the Hondsbossche Zeewering, for example, and the shift from building dikes to widening the rivers. The **fourth factor** is the political constellation, in particular the shift from central management to a regional approach. And finally the **fifth factor** is the rise of the designers and architects. They now also shape the waterworks, together with engineers, even though 70% of the work is under water.'

Marjan den Braber (Royal HaskoningDHV) notes that 'the character of the water system is such that there is not much tolerance for fussing about. These are stubborn systems, which the surroundings have to cope with. This applies to the landscape, and also to the waterworks that are built there.' She mentions the Maori philosophy of the river having its own life, with its own rights. 'We have to start living with the system,' says Den Braber, 'and not the other way around. The Meuse has acquired its rights, and now we need to honour those rights, also in terms of our waterworks.' Tjalle de Haan notes that, in the tradition of the **Tractaet of Dyckagie** by Andries Vierlingh (1507-1579), rules of thumb were used generically. 'In the period when I worked for Rijkswaterstaat, I always encountered custom-made work. What were the specific requirements for this lock, or that pumping station, and what were the solutions? You optimised for that. And the result was always unique.'

'The water system determines what will happen, how you will deal with what happens, and what you have to anticipate,' says Ellen Vreenegoor (Cultural Heritage Agency of the Netherlands). That applies to both bigger and smaller water systems. The regulation of the rivers in the nineteenth century, for instance, is likely to have contributed to the high water emergencies of 1995 and 1998. For the brooks systems, she refers to research into water mill landscapes, commissioned by the Cultural Heritage Agency with the Aa en Maas water board and the Graafse Raam in 2016. In that research, the researchers do not talk about water mills as objects, but rather about the whole of landscape systems in which culture, nature, and heritage are interwoven in the stream valleys. By forming an integrated system, water mills have influenced the hydrological conditions in the stream valleys in many places for centuries. They have contributed to the formation and the habitat differentiation along the streams, in the stream valley plain, and in complete marsh systems and peat formations. Their influence on hydrology was so great that they were decisive for the creation and maintenance of seepage zones and source zones, also at the foot of the valley flanks. In short, water mills had become a guiding factor for the landscape and the ecological development in large parts of the former stream valleys. 'The entire stream system of North Brabant is artificial,' concludes Vreenegoor.

Decisive Factors for Coexistence

In the domain of civil engineering, waterworks are the result of calculations, but in the field of commissioning, contracting, and the social environment, they are also the result of negotiations about things such as the budget and (increasingly) the support base. For example, what does it mean that other organisations (such as nature and environmental groups) have now also become involved in the waterworks? Or organisations that attach more importance to other qualities beyond the hydrological ones?

It is expected that social factors will come to play an even greater role in the decision-making process about the most appropriate hydraulic engineering solutions. In this context, the shift of hydraulic engineering expertise from government to market parties is also relevant. That means that in the coming decades, the significance of waterworks will also evolve: from an institutional oeuvre in the twentieth century to a corporate oeuvre in the twenty-first century.

The influence of the Club of Rome proved to be a crucial social factor in terms of the evolving insights during the construction of the Oosterschelde barrier. Traditional, technically-driven civil engineering art was replaced by technological innovations that gave an answer to the questions posed by social movements. This was a trend reversal that also made it possible to create space for a freer design.
'The Delta Works,' recalls Tjalle de Haan, 'changed the working method: from a traditional approach to scientifically-based construction. Tradition gave way to an experimental approach. The Delta Service was a place of continuous experiments. The height determined for a seawall, for example, was the highest measured water level plus one metre if there was no major wave attack.
The Delta Plan came up with the normative water level: the water level that had a 10^{-4} statistical probability of being exceeded, or 1 in 10,000 years. Or as Minister Algera phrased it at the House of Representatives in 1958, when the Delta Act was being discussed: 'a one percent chance during a human lifetime,' even though that meant you would have to live to be 100. That was the water level that we needed to be able to deal with. This led to much wider, higher dikes of completely different dimensions. And this in turn led to difficulties with the Oosterschelde barrier, the drama at Brakel, and the unrest around Amelisweerd. All at the same time. The outcome was: the goal of safety still had to be achieved, but the landscape, natural, and cultural values also had to be achieved.'
'That's correct,' says Eli Gehasse (Rijkswaterstaat). 'The Delta Works are extremely innovative, but people were also thinking from the point of view of nature in the past as well; there was always an eye for the environment. What you can now see, as the beginning of a shift in thinking, is that in the eternal wave movement between trying to live against the water and trying to live with it, the choice has now been made to live with the water.'

Ties Rijcken (TU Delft) also sees the storm surge barrier at Oosterschelde as a trend reversal. 'Since then, people have been thinking in terms of customisation. The great gesture has disappeared, and now we are really looking at the site, where local architecture is once again given a chance.' He has identified two themes since the fracture in 1986. 'The first is that before this fracture, we invested in waterworks only after a disaster had occurred. From then onwards, model calculations became the guiding factor. And once again with huge investments. It was no longer the disasters, apart from the flooding of 1993 and 1995, that gave rise to new measures. Instead it was legislation, in particular the Wet op de Waterkering (Act on Flood Protection).' As a second theme, Rijcken sees a development that he characterizes as 'ascending the Maslow pyramid.' 'In Maslow's model, the needs of mankind have a hierarchy. To have a higher layer, you also need the layers below it, although the bottom layers do not need to be completely filled first. Now that we have sufficiently filled the 'low'-lying flood risk management layers, a 'low'-lying spatial quality again comes back into the picture. I say 'again,' because with the call that came for 'sobriety and efficiency' during the financial crisis, that degree of attention has disappeared. In other words, where at first it was mainly physiological needs, towards the top it becomes more artistic.' 'That also means,' he says, 'that the higher up on the scale level, the more unique the waterwork becomes, and there is therefore less and less that can be said about it in a generic sense.'

Ellen Vreenegoor points out issues from the design side. 'Long ago,' she says, 'the approach was a fully integrated one. Technologists and artists, such as someone like Jan van Scorel and Andries Vierlingh, always worked together, until specialisms started to appear. This has become a separate factor, regrettably, because finding ways to cooperate with each other was quite commonplace in the past.' What is also unfortunate, according to Marjan den Braber (Royal HaskoningDHV), 'is that we have forgotten the decisive role of individuals. Many 'names' have been lost, even though construction management is so important.'

If that means that form always followed function in the design of waterworks, then Eli Gehasse (Rijkswaterstaat) has her doubts. 'It is a complex game of factors, in which technical knowledge, the available materials, and the location of the waterworks all played a role. It was and still is a process, an organically grown process, in which systems are linked to each other. Rijkswaterstaat has always been designing for a specific function, but also with an eye for aesthetics. Well-known architects, such as Dirk Roosenburg, built soberly, but also efficiently, and certainly not in an ugly way. They went as far as they could within the set budgets. Even if that meant that there had to be some repetition.' Gehasse notes thereby that Rijkswaterstaat's design role has gradually been transferred to the market. 'We no longer write specifications, but instead we outsource things. Nonetheless, we do still set demands on the design, such as the Aesthetic Programme of Requirements for the current Afsluitdijk project.' She observes that 'the challenge for waterworks now is mainly a demand for being sustainable and energy-neutral, and that requires new innovation.'

Towards an Adaptive Approach

Unlike in the twentieth century, with its penchant for the grand gesture (for example the Delta Works), other values are now being chosen for flood risk management in the twenty-first century. Values that arise from new questions, new approaches, new views. Values formed by questions such as: how will we adaptively apply water management? How will we respond to climate change? Which norms, which techniques, and which social considerations will play (important) roles? How will we continue our pathway-dependent tradition, built over many centuries, in the twenty-first century? Can these values jointly tell a new 'grand gesture,' a new story?

An adaptive approach, also for the development of new waterworks, can lay the foundation for this new 'gesture,' this new story. Taking into account uncertainties, and designing on that basis, is the main challenge. For many, that requires a change in thinking. This is a very urgent change because of its currency, as many of the major waterworks are in need of replacement. So now it is no longer a question of developing programmes that will determine the future developments for many decades, but rather thinking in green concepts in which 'blue' and 'green' are united, without immediately linking this to a formal language. And continuing the wave movement that began at the end of the last century: namely, no longer agitating against water, but instead living with it. Above all, this requires working in an integrated way. Designing with an eye for many relevant issues, such as the mixing or separation of salt and fresh water, or taking into account temperature differences, and using instruments that do not kill fish. 'That means that we have to move away from sectoral thinking and working,' notes Marinke Steenhuis (SteenhuisMeurs). 'There has been a huge erosion in terms of knowledge, skills, and creativity, and especially in the interlinking of skills.'

An adaptive approach can bring that coherence back, says Jos van Alphen (Delta Commission staff). 'Approaching things adaptively is the big challenge; the future is going to change. The pace and the direction are unclear and uncertain, and that in itself becomes a trend.' He points out that thinking about flood risk management has become probabilistic. 'The risks and uncertainties are now explicitly taken into account.' He notes that this fact, 'together with an adaptive approach, determines the choice in the replacement challenges for the current waterworks. This means that you build in the fact that you will have to make new interventions in the future. And this means a cultural change, not only for the commissioning bodies, but also for the contractors. Perhaps we should think more in terms of unfinished waterworks that can change according to need. This can lead to an extension of the catalogue, all achieved with semi-finished products.'

Van Alphen points out thereby that the distribution of benefits and burdens is not always done fairly. 'The (re)construction and presence of a waterwork is mainly a burden for the immediate local residents, whereas people who live further afield are not bothered by it, and only enjoy the benefits, the advantages. This is an important point of attention when it comes to local support.

Climate change, with rising sea levels and higher peaks in river water discharges, as well as changes in social needs, particularly the demand for more sustainable energy consumption, all mean that the outdated waterworks not only need to be replaced, but also that their functionality is due for replacement, too.'

Integrated and adaptive water management is necessary to deal with this. Although it provides guiding principles, it does not yet offer tailor-made solutions for the waterworks, nor any tools for implementation and optimization. The traditional focus of addressing sectoral interests will have to shift towards being driven by broader values, and by providing an efficient coordination of these values. 'Multifunctional waterworks offer new opportunities, as long as they are sufficiently approached as an integral task, says Jos van Alphen. 'Think for example of the opportunity that separating water masses offers in terms of energy generation. This is made possible by flows, temperature differences, and the like. This calls for technical innovations, or the reintroduction of past innovations, much in the way that the Hagestein weirs originally served as energy turbines as well.' He also mentions the example of the hydroelectric power station in the Dommel river near Sint-Michielsgestel (pp. 152-153). 'Dommelstroom' is an initiative of local residents, and has since been adopted by Waterschap De Dommel and applied in the design of the hydroelectric power plant and the modifications to the weir in the Dommel. 'And,' Van Alphen continues, 'energy storage is also becoming an option. Today, wind turbines do not always produce energy at the right moments. Water can store that energy if you equip the waterworks for this purpose.'

Path dependence also offers the possibility for adaptive thinking. For a long time, traditions can be held onto until a break, or a 'hiccup' occurs. This can also mean the rediscovery of a lost tradition. The terp phenomenon is a good example of this. And that also applies to the technology of inundation. What possibilities does the Nieuwe Hollandse Waterlinie (New Dutch Waterline) offer as a water storage facility? The ground keeps subsiding, so we are forced to raise it. So you see occasional return of the tidal basin. This also forces us to reflect on how cultural history is taken into account in the further development of the Dutch water system. 'And that,' as Marinke Steenhuis says, 'forces us to think about the opportunities for sustainability. Not only in terms of the reuse of materials, but also in the re-application of materials and techniques that we have unlearned or forgotten. An extension of this idea is new production techniques, such as 3D printing. The current power of computers means that more beautiful designs are possible, such as wonderfully curved pylons, for example.'

Dijkgraaf Roelof Bleker outlines what he sees as the three most important challenges for the water boards in the coming decades: 'Those are the energy transition, the circular economy, and climate adaptation.'He immediately mentions the EQA installation. This project, which was initiated in part by Waterschap Rivierenland (the Rivierenland Water Board), is an all-in-one hydroelectric power station. The first version was installed near Ommeren in September 2016. It is an innovative weir that produces hydropower, and that can also buffer and discharge surface water. This contributes to 'smart drainage,' says Bleker, 'and avoids unnecessary pumping.' A second edition has now also been installed at the Quarles van Ufford pumping station in Alphen aan de Maas. 'The pumping station is self-sufficient in its use of energy,' says Bleker, 'and aspects such as measurements and managing the water level can be done automatically and remotely.'

The EQA installation and weir, made of recycled plastic, are building blocks that allow the water board to contribute to its goal of achieving an energy-neutral way of working by 2030. 'There are endless possibilities to apply this innovation, because there are many locations that have the minimal required gradient of half a metre. That means that there is also a lot of renewable energy to be gained in the water system, and this is one of the most important challenges that I see for the water board in the coming years. It is also important for the water board itself, because we have five million euros in energy costs every year. But of course, it is much more important for society as a whole. In addition to reducing CO2 emissions, saving and generating energy is a great challenge and an opportunity for the water board. This can be achieved with water, via energy generation and buffering, but we also see possibilities in terms of installing solar panels on sites that we manage, such as dikes. Another initiative is the generation of thermal energy at pumping stations. This would allow, for example, a small and unexceptional-looking pumping station near Arnhem to generate energy for 300 local homes. Of course we will not do this ourselves, but rather we will form alliances with energy suppliers. The goal is not to make a profit for ourselves, but to serve the purpose of making the Netherlands energy neutral.'

Climate adaptation is high on Bleker's agenda. 'Together with the municipalities, we are identifying risks, and looking for solutions. Having too much concrete and stone everywhere is also a big problem, not only within cities and the back gardens of their residents, but also in and around nursing homes, for example, where there is very little green space to discover, because everything has been sacrificed to car parks and the like. These areas can be eight to ten degrees warmer than a green environment. That is why we are also looking to work with housing corporations and healthcare institutions. Together with them and the residents, we will sit down at the table and come up with ideas. An ice-skating rink, for example, might suddenly see opportunities for water storage when the area is hit by rainstorms.'

'That is why it is so necessary to involve the local environment in all of this: the individual residents, as well as social organisations and the business community. Together with them, we are looking for opportunities in terms of technical possibilities, but also in terms of improving quality, and that also applies to the design of the waterworks. We have recently been paying less attention to this last point, so we are now tackling it again by setting up new forms of consultation with each project, for example by including a landscape architect. Here as well, we are looking for close cooperation with those who are directly involved, such as the farmers, and in this case the locally very important fruit growers and growers of avenue trees. By providing services to them, such as being prepared to supply extra water when there is a threat of crop damage from night frost, we remain firmly embedded in the local environment. I stress this because I know that the water boards have not only professionalised over time, but also have institutionalised, meaning that for too long, they were focusing too much on themselves.'

Epilogue

In summary, five elements can be identified as determining factors that answer the questions concerning the new requirements and conditions for waterworks in the twenty-first century, and the materialisation of influences such as unstable climate forecasts, the pursuit of ecological and landscape mitigation, and redistributed responsibilities.

First of all is the need for adaptive capacity. The future is uncertain, but it is indeed certain that the major waterworks have reached the technical end of their lifespan. The question then becomes with which perception, and from which perspective, will the new design commissions be drawn up? How different are they from the previous commissions? Back then, design and construction were based on the knowledge of natural forces, and on of one's own technical abilities, based on the scientific knowledge that existed. It was possible to build the Afsluitdijk, for example, due to the knowledge of tidal effects and the experience that was available at that time. All of this is fundamentally different today. The thinking is now dominated by the climate debate, and the threat of rising water levels. The scenarios that are drawn up in this respect take into account possible windfalls and setbacks in these developments, which means that commissioning bodies have to establish a much more flexible demand. They should, for example, limit themselves to designing the shells, which will enable a gradual response to developments. This in turn means that other requirements have to be imposed, for example, on foundations and materials.

Secondly, and following on from the above, a new approach from the designers is necessary. Sustainability, cradle to cradle, and energy-neutral have become essential social requirements, which also need to be applied to the waterworks. This is a challenging task for the designer, who as always gives a cultural interpretation to developments within society. Even now, the designer has to reconcile the spirit of the times with technology.
In addition, and more than ever before, the designer will need to make the waterworks 'readable.' For example by giving, indeed of showing, an answer to the question of 'how does this work?'; this is crucial to maintaining or creating support for the interventions that are needed. And this is also a challenge that is at least as important for the commissioning body: showing what you do, why you do it, and how you do it are clearly necessary for them to maintain their legitimacy.

As a **third element**, an adequate response to the forces of nature is unavoidable. The peaks in the curves of the climate change scenarios, including the rises in sea level, are becoming more and more frequent. Sea levels rose by an average of 1.5 to 2.0 mm per year during the twentieth century, but this amount is expected to accelerate over the coming decades. In the '14 climate scenarios' of the KNMI, a rise in sea level of up to 40 centimetres can be expected by 2050 (compared to the 1981-2010 period). For 2100, an upper value of 100 centimetres has been calculated. But the opposite scenarios also form a challenge: controlling for a lower water level. The response to the forces of nature dominates the Delta Programme, which looks at extreme conditions that include both too much and too little water. A very large part of this programme deals with strategies and measures to prevent drought or mitigate its consequences, for example by making the water level of the IJsselmeer more flexible (which would increase the freshwater buffer), by increasing the extra freshwater supply via the channelized Hollandse IJssel, and by promoting salt-resistant and drought-resistant crops. A researching design strategy is a prerequisite for this. A strategy that embraces the uncertainty. Only by embracing uncertainty can major leaps forward be made.
Another aspect is that the evolving thinking about energy and climate change also lead to a new challenge in terms of the energy that will be required to operate these waterworks. These are currently powered mainly by fossil energy, and that will have to change. Adaptive thinking will therefore also remain essential in our use of energy sources. Being open to new and unknown techniques is a prerequisite for this.

A **fourth element** is the cooperation between the client, usually a government organisation, and the market. When this cooperation becomes based on mutual trust much more than is currently the case, new developments and innovations will also become possible. Mutual trust requires the involvement of market players in the first phase of planning development. An example is Design and Build constructions, in which the market parties are also involved with an EIA application. This can lead, for example, to a joint creation between a water board and a market party. The relationship between the client and contractor must change. This also means that a project no longer has to be parcelled into phases, with each one requiring a (European) call for tenders to be issued. It means working with a consortium from the very beginning, for both the planning and the execution. New forms of contract, and other funding models based on mutual trust, increase the chances for adaptive waterworks. Long-term cooperation makes it possible to adapt, and also makes local and area-specific solutions more promising.

Finally, a **fifth element** is multi-functionality. New waterworks will have to be able to combine water management with other functions, and with other social tasks. For clients such as Rijkswaterstaat and the water boards, this means that they will be acting more as participants in a social process.

Water Envoy Henk Ovink formulates this challenge as follows: 'We live with water, our cities are water cities, and without the polders we would now be living in Germany. We have developed a range of interventions for this purpose, which we are constantly renewing. We have created a huge potpourri of all the possible interventions and solutions. Examples include programmes such as the Delta Works, Room for the River and the Zwakke Schakels (Weak Links), but also the Markerwadden and the Kierbesluit (Kier Decree): it is a bold move to leave open a small gap in the Haringvliet locks when the water level in the Haringvliet is lower than it is at sea. This is important for international fish migration. In this way, migratory fish, such as salmon and sea trout, can cross through the locks towards their spawning grounds, which lie upstream. The opening of the locks also ensures that salt water can flow in, allowing for the salinization of the western part of the Haringvliet.'
'This all proves our adaptive capacity. We again and again demonstrate that we are able to handle a large number of challenges: factors such as ecology, the quality of life, our cultural values, the hard and the soft infrastructure. First we built a dike, the Afsluitdijk, and now we're drilling a hole in it. We are no longer afraid of mixing brackish water and salt water, but instead we look at the ecological quality of the area. And we see what more can be done now. For example, we are looking at the opportunities for 'blue energy,' or the chances for symbiosis and synergy between water quality and flood risk management. We are able to extract energy from the difference in salt concentration between salt water and fresh water; this is being experimented with at the Afsluitdijk. Rijkswaterstaat is also working with provinces, municipalities, and industry on the Tidal Technology Centre in the Grevelingendam, where research is being carried out into how energy can be generated using tidal turbines.'

'You have to dare to look towards the future,' Ovink emphasises, 'and be able to make the connections between economics, ecology, and social and cultural issues. Climate change requires a comprehensive and integrated perspective, and a process of cooperation with everyone. If you know what you want to achieve in the long term, you can work out what steps you need to take today in order to achieve that. Not in a dogmatic way, but an adaptive and flexible way, different every year, learned from the past, and focused on our uncertain yet also inspiring and exciting future.'

'An adaptive capacity is therefore necessary,' he observes. 'The degree of complexity in water management will remain, and will only be getting bigger and bigger. Fortunately, we are increasingly able to understand complexity. We are also using resources to that end. The Delta Programme, for example, has a fund that makes one billion euros available each year for the realisation of projects, but also for research. I see the adaptive capacity as a true journey of discovery that enables incremental innovations.'

'We have never built for eternity, although that may well have been suggested. We always knew that it was good for that moment, but that in the long run, it would have to be tinkered with. We always knew that what is safe today might have to be reassessed the day after tomorrow. Fortunately, we are now able to develop models that we were previously unable to do. We knew the effect of certain forces of nature, but we could not yet manipulate them, let alone influence or control them. Now Deltares has built the longest trough in the world (300 metres), which they can use to simulate any wave pattern that occurs at sea or in rivers. The research being done in this trough can substantially cut costs worldwide.'
Ovink elaborates on the role of the client in relation to the development and maintenance of a support base. 'The (public) commissioning process follows a wave motion, and it is now heading back to the age-old form of the corporation, much in the way that the water boards once started, from the bottom up, as a local community.'

'Rijkswaterstaat is also choosing for this same approach. There as well, the old top-down approach is now making way for development starting from the bottom up. It is choosing for local embedding, for example in Nijmegen, where the municipality was given the lead in the Room for the Waal programme. The municipality assembled coalition partners so as to be stronger and smarter, and of course because it did not itself have all that knowledge and skills in-house. With whom, and how, are we going to approach this for the best result? That was the guiding question. Organizing this in a collective way creates ownership, and it becomes a project of many. That makes it stronger. On your own, you can never come up with as many ideas as you can together with others.'

'Such a broadly supported approach keeps a project away from populism, and boosts confidence: 'we can achieve something together.' This also raises awareness, and gives ownership, also to the process itself, and strengthens the possibilities of the realization.'

'This is actually the core of our polder model. It is not all about negotiations, in which you always lose something (which you sometimes take into account in advance, and other times meet with dismay), but rather about cooperation. Collaboration means investing in each other, so that something is created in common: always involving something provided by someone else, as well as something from yourself. In other words interaction, nourished by curiosity, by innovative power. Inclusion and transparency are key concepts in this cooperation. If the process is transparent, then failures – an inevitability in experiments – offer an opportunity for evaluation and learning. As a company or as a government, this enables you to build up crucial knowledge and capacity. To innovate, you have to be a bit rebellious, to go off the beaten track. For that, we need that culture of openness, and we very much need that rebelliousness in the challenges of tomorrow. They are large and complex, far-reaching for everything and everyone, and will require all of our creativity and power of collaboration. If we want to change the world – and we have to, we have no choice – then that can only be achieved with everything and everyone. And we have already known that for centuries.'

Acknowledgements

About seven years ago, the idea arose to make a book about the waterworks in the Netherlands.[IV] A book about so-called 'wet works of art': structures that have a button or a lever attached that can affect the flow and/or direction of the water.
After making an initial global inventory of the various types of waterworks – locks, weirs, and pumping stations – I started photographing. In 2010, I took pictures of the locks at IJmuiden, as well as the weir and lock complex at Grave. These formed the substantive and visual starting point for this photo project.
Most of the photography that I found during the course of my research into the waterworks depicts the waterworks themselves, especially their technology, even though to me it seemed that their location and embedding in the landscape were at least as important. Photographically, that was also much more in line with my earlier work, in which I photographed the city as a landscape. The inspiration for this was a photographic movement that originated in America in the 1970s: New Topographics. Photography of the landscape with a plate camera, as it was done in the earliest days of photography. Photography that did not want to be spectacular, but merely to show what is there. This approach was very much in line with my preference to show the city in all its baldness. To introduce a kind of superficiality in my work: '… a description of the surface is often much more interesting than that of depth, of that which lies beneath the surface…'[V]

Early in the process, I decided not to create the photos in colour, but rather in black and white, which allows the photographs to take on a level of abstraction in which time seems to have stood still. It is as if all the photographs come from the same era, and it imbues the waterworks, regardless of their age, with a uniform appearance.

The initial results prompted the publisher nai010 to take part in the process. This also applied to a number of other people who joined the team of contributors that we assembled. I had asked them because of their specific knowledge in different areas (policy, design, and technology). I didn't want to make a photo book, but rather a book with solid photography that was embedded in a substantive story. I wanted to bring a number of things together in this document: a brief historical overview of the creation of the water boards and other policy-making authorities; the technology and operation of the various waterworks; and a substantive story about the current state of affairs, both in terms of policy as well as in terms of technical thinking and ability. The idea arose to organise two expert meetings with professionals from the field, both from the policy side and from the construction side. These meetings form the basis for the essay.

Of course, I was nether able nor willing to photograph every waterwork; this book is not an exhaustive picture of what is happening in the Netherlands in terms of waterworks. Instead, it is a personal choice, supported by the editors, in which large and small waterworks throughout the Netherlands have been portrayed. With this book, I wanted to show the necessity, the beauty, and the almost self-evident nature of the waterworks in our country. The Netherlands cannot do without them, and we no longer stop to think about just how much these waterworks have become part of our essence.

Finally, I would like to thank everyone who has worked on this book. Many thanks go out to the editorial staff: Inge Bobbink, Eric Luiten, and Lodewijk van Nieuwenhuijze. And to Theo van Oeffelt, who wrote the two long essays, and who was also a member of the editorial staff, and Bernard Hulsman, who wrote pointed reflections on many of the photographed projects. Finally, Emma Ottevanger created beautiful illustrations that outline the functioning and the location of the waterworks. Thanks as well to Yvonne van Versendaal for the clear design of the book.

IV. In collaboration with Egbert Koster.
V. Italo Calvino in an interview in the Dutch newspaper NRC, 18-06-1993

Participants in Expert Meeting (I) and/or (II)
Jos van Alphen, Delta Commissioner (I and II)
Inge Bobbink, TU Delft (I and II)
Marjan den Braber, Royal HaskoningDHV (I and II)
Eli Gehasse, Rijkswaterstaat (I)
Tjalle de Haan, Association of Waterbuilders (I)
Nardo Hoogendijk, Hollandia Infra b.v. (I)
Luuk Kramer, photographer (I and II)
Eric Luiten, landscape architect (I and II)
Lodewijk van Nieuwenhuijze, H+N+S Landscape Architects (I and II)
Theo van Oeffelt, author (I and II)
Ties Rijcken, TU Delft (I)
Jasper Schilder, Dura Vermeer (II)
Marinke Steenhuis, SteenhuisMeurs (II)
Berno Strootman, Rijksadviseur voor de Fysieke Leefomgeving (I)
Hein Versteegen, Dura Vermeer (I)
Ellen Vreenegoor, Cultural Heritage Agency of the Netherlands (I and II)

Interviews
Roelof Bleker, Dutch Water Authority Rivierenland
Henk Ovink, Water Envoy
Ellen Vreenegoor, Cultural Heritage Agency of the Netherlands

Literature
Abrahamse, J.E. & Zee, A. van der. (2016). **De Atlas van de Schie**. Bussum: Thoth
Aten, D. et al. (2007). **Hollanders en het water**. Hilversum: Verloren
Bendeler, G. et al. (1998) **Nat & Droog**. Amsterdam: Architectura & Natura
Brinke, W. ten. (2007). **Land in zicht**. Amsterdam: Veen
Boosma. K, et al. (2007). **Bouwen in Nederland**. Zwolle: Waanders uitgevers
Mars, H. de, & Rijsselt, E. van, (2013). **Watermolenlandschappen**. Roermond, Huis voor de kunsten
Meyer, H. (2016). **De staat van de delta**. Nijmegen: Vantilt
Pols, L. & Strootman, B. (1999). **De landschapsstad**. Den Haag: NIROV
Reh, W., Steenbergen, C., Aten, D. (2005). **Zee van land**. Wormer: Stichting Uitgeverij Noord-Holland
Cultural Heritage Agency of the Netherlands (undated). **Handreiking Water erfgoed en ruimte**. Amersfoort: RCE
Schmeink, T. (2000). **Nederland waterstaat**. Amsterdam: Stichting Open Monumentendag
Steenhuis, M. at al. (2016). **De Deltawerken**. Rotterdam: nai010 publishers
Woud, A. van der, (1987). **Het lege land**. Amsterdam: Atlas

Credits

Concept, Photography
Luuk Kramer

Texts
Theo van Oeffelt and Bernard Hulsman, with Inge Bobbink and Ellen Vreenegoor contributing

Translation
Douglas Heingartner

Editing
Inge Bobbink, Eric Luiten and Lodewijk van Nieuwenhuijze

Copy Editing and Correction
Brenda Heeringa

Design
Yvonne van Versendaal

Lithography and Printing
Wilco Art Books, Amersfoort

Publisher
Eelco van Welie, nai010 uitgevers, Rotterdam

This publication was made possible with generous support of BPD Cultuurfonds, Creative Industries Fund NL, Cultural Heritage Agency, Rijkswaterstaat and Royal HaskoningDHV

bpd creating living environments

creative industries fund NL

Rijksdienst voor het Cultureel Erfgoed
Ministerie van Onderwijs, Cultuur en Wetenschap

Rijkswaterstaat
Ministry of Infrastructure and the Environment

Royal HaskoningDHV
Enhancing Society Together

nai010 publishers is an internationally orientated publisher specialized in developing, producing and distributing publications on architecture, visual arts and related disciplines. www.nai010.com

nai010 books are available internationally at selected bookstores and from the following distribution partners: North, Central and South America - Artbook | D.A.P., New York, USA, dap@dapinc.com
Rest of the world - Idea Books, Amsterdam, the Netherlands, idea@ideabooks.nl

For general questions, please contact nai010 publishers directly at sales@nai010.com or visit our website www.nai010.com for further information.

ISBN 978-94-6208-386-8
Printed and bound in the Netherlands